# FOOD SYSTEMS FAILURE

This book provides a critical assessment of the contemporary global food system in light of the heightening food crisis, as evidence of its failure to achieve food security for the world's population. A key aspect of this failure is identified in the neoliberal strategies which emphasize industrial efficiencies, commodity production and free trade ideologies that underlie agricultural and food policies in what are frequently referred to as 'developed countries'. The book examines both the contradictions in the global food system as well as the implications of existing ideologies of production associated with commodity industrial agriculture using evidence from relevant international case studies. The first section presents the context of the food crisis with contributions from leading international academics and food policy activists, including climate scientists, ecologists and social scientists. These contributions identify current contradictions in policy and practice that impede solutions to the food crisis. Set within this context, the second section assesses current conditions in the global food system, including economic viability, sustainability and productivity. Case study analyses of regions exposed to neoliberal policy at the production end of the system provide insights into both current challenges to feeding the world, as well as alternative strategies for creating a more just and moral food system.

**Christopher Rosin** is a Research Fellow and Deputy Director with the Centre for Sustainability: Agriculture, Food, Energy, Environment at the University of Otago in Dunedin, New Zealand. His primary research interest is in the social negotiation of sustainability in the agricultural sector.

**Paul Stock** is a Lecturer of Sociology and Research Fellow with the Centre for Sustainability: Agriculture, Food, Energy, Environment at the University of Otago in Dunedin, New Zealand. Paul's current research interests include the social and cultural aspects of agriculture, the intersection of morality and the environment, and the Catholic Worker movement.

**Hugh Campbell** is Professor of Sociology and former Director of the Centre for Sustainability: Agriculture, Food, Energy, Environment at the University of Otago in Dunedin, New Zealand. His research interests include rural sociology, sustainable agriculture, neoliberalism and agrifood governance, food waste, masculinity and rural gender.

# FOOD SYSTEMS FAILURE

## The Global Food Crisis and the Future of Agriculture

*Edited by*
*Christopher Rosin*
*Paul Stock*
*Hugh Campbell*

Routledge
Taylor & Francis Group

LONDON AND NEW YORK

earthscan
from Routledge

First published 2012
By Earthscan from Routledge

First published in paperback 2014
by Routledge
2 Park Square, Milton Park, Abingdon, Oxon OX14 4RN

And by Routledge
711 Third Avenue, New York, NY 10017

*Routledge is an imprint of the Taylor & Francis Group, an informa business*

© 2014 selection and editorial material, Christopher Rosin, Paul Stock and
Hugh Campbell; individual chapters, the contributors

*British Library Cataloguing-in-Publication Data*
A catalogue record for this book is available from the British Library

*Library of Congress Cataloging-in-Publication Data*
Food systems failure : the global food crisis and the future of agriculture /
[editors] Christopher Rosin, Paul Stock, Hugh Campbell.
p. cm.
1. Food supply. 2. Food industry and trade. 3. Hunger.
4. Food security. I. Rosin, Christopher J. (Christopher John)
II. Stock, Paul V., III. Campbell, Hugh, 1964-
HD9000.5F599 2011
338.1—dc23
2011021896

ISBN: 978-1-84971-229-3 (hbk)
ISBN: 978-0-415-71260-6 (pbk)
ISBN: 978-1-84977-682-0 (ebk)

Typeset in Times New Roman
by Bookcraft Ltd, Stoud, Gloucestershire

Printed and bound by CPI Group (UK) Ltd, Croydon, CR0 4YY

# CONTENTS

# LIST OF FIGURES AND TABLES

## Figures

## Tables

# CONTRIBUTORS

**Bustanul Arifin** is Professor of Agricultural Economics in the University of Lampung and Professorial Fellow in the International Center for Applied Finance and Economics at Bogor Agricultural University (IPB). He has published 36 books and over 80 journal articles and book chapters. He is currently a member of the Food Security Council, National Innovation Council, and the Advisory Team for International Trade Negotiations.

**Colin D. Butler** is a research academic at the Australian National University, and also medical director of BODHI, an NGO he co-founded in 1989. BODHI has projects with poor and marginalized people in seven Asian countries. In 2009, Colin was named as one of the 100 "doctors for the planet".

**Hugh Campbell** is Professor of Sociology and former Director of the Centre for the Study of Agriculture, Food and Environment (CSAFE) at the University of Otago, New Zealand. His research interests include rural sociology, sustainable agriculture, neoliberalism and agri-food governance, food waste, masculinity and rural gender.

**Michael Carolan** is an Associate Professor of Sociology at Colorado State University. His areas of expertise include environmental sociology, food and agricultural policy, and social theory. He is author of the recently published books *The Real Cost of Cheap Food* (Earthscan, 2011) and *Embodied Food Politics* (Ashgate, 2011).

**Jane Dixon** is Senior Fellow at the Australian National University. Her research on the public health implications of food system transitions takes place in Australia and Thailand.

**Ian Gray** is Adjunct Associate Professor at Charles Sturt University at Wagga Wagga, NSW. His research interests cover the sociology of community, rural sociology, local government, urban society, transportation and the environment. He has conducted numerous research studies in rural Australia including investigations of drought, farming practices, small-town sustainability and attitudes to regionalism.

**Naomi Hansar** received first class honours for her thesis 'Community Supported Agriculture: An Expression of Environmental Citizenship?' at The University of Queensland in 2009, under the supervision of Professor Geoffrey Lawrence. During 2010 and 2011 she has been Coordinator of the Food Security Focal Area for the Global Change Institute at the University of Queensland.

**Doug Hill** is Senior Lecturer of Geography at the University of Otago, New Zealand. His research has focused on comparative political economy in South Asia, rural development, port sector and maritime trade; common property issues, especially forests; and transboundary issues related to water resources.

**Geoffrey Lawrence** is Professor of Sociology, Co-Leader of the Food Security Focal Area for the Global Change Institute, University of Queensland, and a Fellow of the Academy of Social Sciences in Australia. His career includes over $10 million in grants and 25 books, most recently *Supermarkets and Agri-food Supply Chains* (Edward Elgar, 2007) and *Food Security, Nutrition and Sustainability* (Earthscan, 2010).

**Kristen Lyons** is Senior Lecturer in the School of Social Science at the University of Queensland, researching topics related to the sociology of food, agriculture and the environment. She has been involved in food-related campaigns with Friends of the Earth Australia. She co-edited *Food Security, Nutrition and Sustainability* (Earthscan, 2010).

**Claire Mahon** is an international human rights lawyer based in Geneva. She was advisor to the former UN Special Rapporteur on the Right to Food, Jean Ziegler, and Special Advisor to former UN High Commissioner for Human Rights, Mrs Mary Robinson. She is the co-author of The Right to Food: Lessons Learned.

**Philip McMichael** is Professor of Development Sociology, Cornell University. He authored *Settlers and the Agrarian Question* (1984) and *Development and Social Change* (2008), and edited *Contesting Development: Critical Struggles for Social Change* (2010). He has worked with FAO, UNRISD, Vía Campesina and the IPC for Food Sovereignty and is researching the land-grab question.

**Jeff Neilson** is a Lecturer in Geography at the University of Sydney, where he teaches on economic geography, natural resource management and rural development in South East Asia. His research interests include the environmental and livelihood implications of smallholder engagement with global value chains, for agricultural commodities such as coffee and cocoa.

**Jules Pretty** is Professor of Environment and Society at the University of Essex. His 18 books include *This Luminous Coast* (2011), *Nature and Culture* (2010), *The Earth Only Endures* (2007), and *Agri-Culture* (2002). He has received an OBE (2006) for services to sustainable agriculture and an honorary degree (Ohio State University, 2009). More details can be found at www.julespretty.com.

**Bill Pritchard** is Associate Professor in Economic Geography at the University of Sydney, with research interests in the links between globalization, agri-food industries and food security. He has published widely on topics including agri-food value chains, regional development issues in Australia, and rural restructuring in India.

**Carol Richards** is a Postdoctoral Research Fellow in the School of Social Science at the University of Queensland. She has published in the areas on food, farming and natural resource management – with a recent focus on mass food retailing and retailer power. She is the current Convenor of the Australasian Agri-food Research Network.

**Christopher Rosin** is a Research Fellow at the Centre for Sustainability: Agriculture, Food, Energy, Environment, University of Otago. As a member of the transdisciplinary ARGOS research team, his research involves the interrogation of agricultural sustainability including farmer response to climate change and the moral justification of farming practice.

**Kiah Smith** is a Research Analyst at the United Nations Research Institute for Social Development (Geneva) with a focus on corporate social responsibility and multi-stakeholder regulation, green economy, and alter-globalization movements. She holds a PhD in Environment and Development Sociology (University of Queensland) and maintains research interests in global food security, gendered political economy and qualitative/participatory research methods.

**Paul Stock** is a Lecturer in Sociology and Research Fellow with the Centre for Sustainability: Agriculture, Food, Energy, Environment at the University of Otago in Dunedin, New Zealand. Paul's current research interests include the social and cultural aspects of agriculture, the intersection of morality and the environment, and the Catholic Worker movement.

**Alec Thornton** is a Lecturer in Development Geography at the University of New South Wales in Canberra, Australia. His research interests include urban agriculture and urban ecosystems, religion and civil society, capacity building and community-based development in developing countries.

**Navé Wald** is a PhD candidate in the Department of Geography, University of Otago, New Zealand. His work focuses on peasant and indigenous organizations in Northwest Argentina and their discourses, challenges and praxis, examining them as particular models of development.

**Robert T. Watson** is Professor of Environmental Sciences and Director of Strategic Investment at the Tyndall Institute (University of East Anglia). He also serves as the Chief Scientific Advisor to the Department for Environment, Food and Rural Affairs (Defra) in the UK, and has served as chair of the IAASTD (current) and the IPCC (former).

# PROLOGUE: FOOD SECURITY – NOW IS THE FUTURE

## Robert T. Watson

There is widespread realization that, despite significant scientific and technological achievements resulting in our ability to increase agricultural productivity to meet growing demand, there have been unintended social and ecological consequences of these achievements.

It is undeniable that over the past century, agricultural science and new technologies have boosted production, with enormous gains in yields and reductions in the price of food. But these benefits have been unevenly distributed, e.g. today over one billion people still go to bed undernourished every night, especially in parts of sub-Saharan Africa and South East Asia – in the last couple of years there have been an additional 100-150 million people affected by the increase in food prices and the global economic downturn. Primarily this is a problem of distribution and local production, but whatever the causes, solutions are going to be increasingly difficult. At the same time, the focus on increased production has contributed to an increase in greenhouse gas emissions and, in some parts of the world, land and water degradation due to unsustainable intensification practices, coupled with the loss of biodiversity (genetic, species and ecosystem level) and ecosystem services due to extensification.

In coming decades we need to respond to a doubling of food demand, meet food safety standards, enhance rural livelihoods and stimulate economic growth in an environmentally and socially sustainable manner. All of this at a time when the rate of increase in productivity per hectare for most cereals is decreasing, and when there will be less labour in many developing countries as a result of HIV/ AIDs and other endemic diseases (e.g. malaria in Africa); when competition from other sectors will make water even more scarce; when there will be less arable land due to soil degradation and competition from biofuels; when biodiversity is being lost at the genetic, species and ecosystem level; and when the climate will be changing, giving us higher temperatures, changing rainfall and more frequent floods and droughts.

In the last couple of years there has been a period of increased food prices. The underlying causes of these increases are inevitably complex and include factors such as increased demand from rapidly growing economies (especially China); poor harvests due to an increasingly variable climate (e.g. the Australian drought); the use of food crops for biofuels (e.g. maize for bioethanol); higher energy and fertilizer prices; low food stocks per capita; export restrictions on agricultural products from a number of significant exporters to protect domestic consumers (e.g. Argentina, India and Ukraine); and speculation on the commodity futures market. In addition, many developed-country agricultural import tariffs and export subsidies distort global markets – in some cases depressing world prices, for example via subsidized 'dumping' (making local production difficult in developing countries), and increasing global prices by inflating OECD prices.

A key question is whether these price increases are a momentary blip – the result of an unfortunate series of events – or are they a harbinger of the future? Some factors impacting food prices are shorter term than others. For example, the effects of adverse weather conditions tend to be relatively short-lived, but are also recurrent. There is already evidence that the current high prices are stimulating increased production, but it may take a number of years to rebuild stocks to levels that markets are comfortable with. Longer-term issues include the future cost of energy and impact of global warming, which may give rise to more enduring climate change and more frequent occurrences of extreme weather events – leading in turn to potentially greater agricultural price variability in future. And if they are more than a blip, what else do we need to know if we are to provide sustainable, nutritious and affordable food for the world in an environmentally and socially sustainable manner?

Meeting the goal of affordable nutritious food for all in an environmentally sustainable manner is achievable, but not through current agricultural 'business-as-usual'. Instead, if a large part of the world is to be prevented from going hungry in the twenty-first century, we need nothing short of a new 'agricultural revolution', with a more rational use of scarce land and water resources and an equitable trade regime, as well as widespread recognition of, and action on, climate change. We also need to acknowledge that in this changing world we need new tools, which means increased investments in agricultural knowledge, science and technology. The new agricultural revolution needed to meet this challenge will require a fundamental rethink of the role of such agricultural knowledge, science and technology. Agriculture can no longer be thought of as production alone: instead, the inescapable interconnectedness of agriculture's different economic, social and environmental roles and functions must be explicitly recognized.

Thankfully, many of the technologies and practices we need to meet the challenge of sustainable agriculture already exist. For instance, we know how to manage soil and water effectively to increase water retention and decrease erosion; we already have access to microbiological techniques to suppress diseases in soils; and conventional biotechnology (plant breeding) can help us produce improved crop varieties. But climate change and new and emerging animal diseases are simultaneously throwing up problems previously unconsidered, and which will need advances in agricultural knowledge, science and technology to address. In addition, we need to use technologies that already exist to reduce post-harvest loss and improve food safety. As appropriate, we need to integrate local and traditional knowledge with formal knowledge, ensuring that the needs of the small-scale farmer are addressed.

Climate change has the potential to irreversibly damage the natural resource base on which agriculture depends and, in general, adversely affects agricultural productivity. While moderate increases in temperature can have small beneficial effects on crop yields in mid- to high-latitudes, in low-latitudes even moderate temperature increases are likely to have negative yield effects. Water scarcity and the timing of availability will increasingly constrain production, and it will be critical to take a new look at water storage to cope with more extreme precipitation events, higher intra- and inter-seasonal variations (floods and droughts), and increased evapotranspiration. Climate change is already affecting – and is likely to increase – invasive species, pests and disease vectors all adversely affecting agricultural productivity. Advances in agricultural knowledge, science and technology will be required to develop improved crop traits, for example temperature, drought, pest and salt tolerance. In addition, it will be critical to reduce greenhouse gas emissions from the agricultural sector – methane from livestock and rice and nitrous oxide from the use of fertilizers.

While biofuels can offer potential benefits over the rising costs of fossil fuels, energy security issues, reducing greenhouse gas emissions and rural economies, the IAASTD (2009) concluded that the production of first-generation biofuels, which are predominantly produced from agricultural crops (e.g. bioethanol from maize), can raise food prices and reduce our ability to alleviate hunger. There is also considerable debate over the environmental impact of biofuels, including their greenhouse gas emissions and their impact on biodiversity, soils and water. Increased public and private investments are needed to develop next-generation biofuels, such as cellulosic ethanol and biomass-to-liquids technologies, so that cheaper and more abundant feedstocks can be converted into biofuels, potentially reducing the demands for agricultural land.

Opening national agricultural markets to international competition can offer economic benefits but, without basic national institutions and infrastructure being in place, can also lead to long term negative effects on poverty alleviation, food security and the environment. Therefore, trade policy reform that provides a more equitable global trading system can help make small-scale farmers profitable and enhance the ability of developing countries to achieve food security while ensuring environmental sustainability, for instance by eliminating OECD production subsidies, eliminating tariff escalation on processed products in both developed and developing countries, and recognizing the special needs of the least developed countries through non-reciprocal market access.

Currently, the most contentious issue in agricultural science is the use of recombinant DNA techniques to produce transgenic products, primarily because there is not yet widespread agreement on the environmental, human health and economic risks and benefits of such products. Many believe that less technology and intervention is the answer. But, against a backdrop of a changing climate and the threat of even larger parts of the world going hungry, it is clear that integrated advances in biotechnology, nanotechnology, remote sensing and communication technology, for instance, will be important in providing opportunities for more resource-efficient and site-specific agriculture. For any technology it will be critical to assess the risks and benefits on a case-by-case basis.

Today's hunger problems can unarguably be addressed with appropriate use of current technologies, emphasizing agro-ecological practices (e.g. no/low till, integrated pest management and integrated natural resource management), coupled with decreased post-harvest losses, coupled more broadly with trade reform and rural development. Small-scale farmers need access to the best seeds, to financing and to markets; we need to create opportunities for innovation and entrepreneurship, and to invest in science and technology and extension services to meet their needs. We also need to provide payments to farmers for maintaining and enhancing ecosystem services, and to recognize the important role of women – empowering them through education, access to financing, and property rights. Specifically, our ability to produce affordable nutritious food accessible to everybody in the future will mean addressing several of the drivers of the current increase in food prices. We will need to decrease the vulnerability of agricultural productivity to projected changes in climate; develop the next generation of biofuels; and transform the trade system to benefit the small-scale farmer.

Meeting the goal of affordable nutritious food for all in an environmentally sustainable manner is achievable. The future is not pre-ordained, but is in our collective hands. While we can build upon our successes, we must also recognize that an extrapolation of business-as-usual will not suffice. Instead, we need to be bold enough to rethink agriculture. Most importantly, if we are to help today's and tomorrow's poor and disadvantaged, we need to acknowledge that the time to act is now.

## References

IAASTD (International Assessment of Agricultural Knowledge, Science, and Technology for Development) (2009) '*Agriculture at a Crossroads: The Global Report*', Island Press, Washington, DC

# INTRODUCTION: SHOCKING THE GLOBAL FOOD SYSTEM

*Christopher Rosin, Paul Stock and Hugh Campbell*

The spike in global food prices, peaking in 2008, is commonly recognized as a shock to the global food system by media, governments and international organizations, and academics. The broad resonance of this shock reflects the challenge that localized food scarcity, and subsequent popular protest (often referred to as 'food riots') in Africa, Asia, South America and the Caribbean, posed to a shared sense of progress – and some would argue complacency – toward meeting the world's food demands. After all, one of the central conceits of the last years of the twentieth century held that, even if hunger was still with us, it was only a matter of time before a combination of trade liberalization, expanding food markets, new technologies, post-Cold War political stability and economic growth in the Developing World rendered the spectre of a hungry world obsolete. Faith in the potential of these dynamics was underpinned by the consistent downward trend in the cost of food commodities in the second half of the century. The real threat and shock value of the food crisis was most evident, however, in the erosion of cheap food as a pillar of global food security. Clearly, an ability to feed the global population was proving to be less certain and hunger on a large scale was still a reality.

Understanding the causes of the global food crisis has proved elusive, not for any lack of potential contributing factors, but for the absolute abundance of competing explanations. The crisis led food system analysts to re-examine the relationships between food production, the looming end of plentiful petroleum-based energy (also known as 'peak oil'), a rapidly changing global climate and volatility in financial markets. While both inequality in distribution of, and access to, food, as well as expanding and more intensive consumption patterns still threaten global food security, the competing demands for agricultural products from a burgeoning biofuels industry was quickly identified as a villain of the food crisis. Similarly, the impact on global food supplies of crop failures caused by climatic extremes in Australia, Vietnam and Russia, suggested that food scarcity would be more common in a future affected by anthropogenic climate change. These factors were further compounded by the economic uncertainty linked to the financial crisis in the US housing market. Investors looking for more secure investment options were blamed for accelerating price increases for food commodities by 'bidding up' the value of grain futures. Some talked of a 'perfect storm' which might, after all, prove to be a 'blip' before things return to business-as-usual. Others are less convinced, viewing 2008 as a harbinger of an increasingly dire global food situation.

One result of the sense of crisis in the global food system was a proliferation of conferences and meetings intended both to develop explanations of the crisis, and to propose solutions or pathways for the future. This book is a development of one such conference at the University of Otago in New Zealand – the 44th Otago Foreign Policy School: Dimensions of the Global Food Crisis, 26–28 July 2009.[1] The critical nature of the situation was evident in the ability of a relatively small, regional conference to attract the interest and contribution of recognized international experts such as Robert Watson, Jules Pretty, Jean Ziegler (represented by Claire Mahon) and Tim Lang. While the Otago Foreign Policy School is intended to focus on issues of contemporary importance as they relate to New Zealand foreign policy, it became very evident that the role of a small nation in the South Pacific extended well beyond its contribution to the global supply of milk, meat, fruit and wine. Thus, the book has evolved as a broader examination of the global food system using the food crisis as a vantage point from which to scrutinize the social, moral and environmental landscape of food provision. While there may not be complete agreement amongst the contributors as to the exact nature of the crisis (as a temporary 'blip', as a reflection of established structural conditions, or as a harbinger of greater challenges in the future), all of them recognize the significance of the event as a feature of the vulnerabilities and uncertainties surrounding the imperative to feed the world. As a whole, the contributions also indicate the need for a fundamental shift in orientation and emphasis necessary for developing a more resilient global food system.

## The food crisis as an 'event'

The common representation of the popular response to rapidly escalating food commodity prices from 2006 to 2008 was that of a crisis event. In this manner, the food crisis invaded our consciousness, newspaper front pages and television screens as a lightning flash that disrupted our sense of complacency and achievement in regard to global hunger. Prior to this event, a general sense had emerged that, at a global scale, hunger was to be associated with environmental disaster, civil and ethnic conflicts, or the actions of despotic rulers. Our capacity to produce sufficient food as a global community and the commitment of 'responsible' governments to meet the goals documented in the Millennium Development Goals and the Universal Declaration of Human Rights surely indicated that we were on the road to solving global hunger. In specific situations of need, aid and development interventions led by individual nations provided an answer. Otherwise, new institutions like the World Trade Organization (WTO) helped expand free trade, providing the central mechanism for 'solving' world hunger. The outbreak of massive WTO protests in Seattle and Cancun during the collapse of the most recent round of world trade talks was initially interpreted as evidence of failure of the ability of the world's nations to agree on world trade policies, particularly around food. After 2008, the political courage expressed as resistance to further trade liberalization became aligned with a wider sense of the failure of market solutions to pressing issues like hunger.

It is worth noting that the upsurge in food prices and the fluctuating numbers of chronically hungry, vacillating somewhere between 800,000,000 and 1,200,000,000 people, did not constitute a 'crisis'. Volatility and the tendency to spike in response to certain market changes are, after all, normal and acceptable features of prices in commodity markets. The reference to crisis only entered the dialogue when people began to protest and threaten political stability. Already, in late 2007, the rumblings of discontent over rising food prices surfaced as public protest in Uzbekistan and Mauritania. A continued sharp rise in food commodity prices the following year evoked similar protests in over 30 countries, ranging from Indonesia and Pakistan to numerous African states and to Latin America and the Caribbean.

The most prominent of these events attracted greater international attention as a result of the associated threats to the stability of governments and the sometimes brutal actions that were employed to suppress the protests. Numerous international experts, including those at the Food and Agriculture Organization (FAO) of the United Nations, and the World Bank, as well as independent scholars such as Jeffrey Sachs, argued for a rapid response to the food crisis precisely because of its potential to destabilize governments in the worst affected countries. Some governments responded to such perceived threats with force. Protesters were arrested and jailed (for example, in Morocco, Burkina Faso and Mauritania) and even killed (in Cameroon and India). Some of the more lasting images of the food crisis are those of Haitian street demonstrations that resulted in the resignation of the country's prime minister, and those of rioting and repression in West Africa and India. Media representations of the crisis were, to a large extent, defined less by the hunger it engendered than by its political implications.

Additional global concerns were raised as governments responded to the popular uprisings by introducing economic policies – subsidies, wage increases and export restrictions on food – that contradicted 'best practice' established by World Bank and International Monetary Fund (IMF) economists. For some countries, this involved increasing government spending to improve access to food; for others, there was a retreat from the free-market principles instilled by the international community. In these cases, such responses demonstrated the potential for high food prices and hunger to unsettle the seemingly well-structured and secure process of incorporating developing countries within broader understandings of economic normalcy and best practice. Responding to a crisis, the World Bank and IMF recommended stop-gap solutions – a response intended to emphasize the necessarily temporary nature of any deviation from the economic policies and structures that they were promoting. The food crisis also exposed the error of a

3

false reliance on 'cheap food' as the foundation for continued economic growth. The assumption that inexpensive food would ensure the low cost of labour in developing economies was a further element of the complacency that had taken hold within the global food system.

Our strong conviction that food security was no longer an issue created the need for a cause, for some villainous agent on which to pin this calamity. The list of suspects was substantial, providing targets to appeal to a variety of political and economic perspectives. The first set of suspects involved the demand side of the global food system, beginning with the newly emerging demand for grain to feed biofuel production as a solution to dwindling oil reserves and climate change. Because the introduction of incentives to promote biofuel coincided with the surge in food commodity prices, they were readily branded as a cause of the crisis. Other demand side factors identified as contributing to the crisis included rapid population growth and the escalating middle-class demands for less efficiently produced foods such as meat and dairy products, with China and India noted as specific examples. The impact of population growth was further exacerbated by accelerating urbanization and the increasing demand for imported food commodities in developing countries. These latter processes were the focus of assessments by neo-Malthusian and 'limits to growth' proponents including Paul Ehrlich of Stanford University and Lester Brown of the World Resources Institute.

Several food systems experts and organizations also emphasized the role of supply side effects on food availability and prices. In some cases, this argument coincided with concerns about global climate change as extreme weather events in Australia, Russia and elsewhere reduced the harvest of staple grains. Others, including Julian Cribb (2010), saw evidence that the lack of investment in agricultural science had resulted in diminishing productivity gains and increased environmental degradation. The World Bank and the IMF, meanwhile, promised greater investment emphasis on agricultural infrastructure in developing countries as a means to encourage increased production through access to markets. Lead economists did, however, acknowledge that it would be naive to expect an instantaneous response to, and benefit from, such policies.

Later explanations of the food crisis accounted for the diverse factors distorting food commodity prices as a perfect storm; in other words, the crisis was not so much the result of a single disrupting influence but rather, a result of the compounding effects of coincidental factors. Early references to a perfect storm came from those faced with disgruntled populations such as President Elías Antonio Saca of El Salvador. The metaphor was soon taken up by representatives of international organizations (including the UN World Food Programme) in order to emphasize the gravity of the situation and the improbability of a simple and fast resolution. The perfect storm metaphor implied, however, that the global food system functions well under normal conditions and that the real failure during the crisis was the unfortunate and ill-timed simultaneous pressures on food commodity supplies and prices. Thus, rather than acknowledging a fundamental flaw in the global food system, the crisis was seen as a situation that could be alleviated without revising existing approaches to food security.

More recently, both the media, and global food system experts have begun to reassess the food crisis and its causes as an event with identifiable boundaries. These analyses generally argue that one or more of the suspected causes of the crises may have been overemphasized at the time. Some of these reassessments, for example by the International Food Policy Research Institute (IFPRI), refer to comparisons to the earlier food crisis in 1972–74, resulting from a rapid increase in oil prices and pressure on commodity stocks due to increased purchasing from the Soviet Union and Eastern Europe. The general conclusion of such analyses is that the contributing factors in 2007–08 were substantially different from those in the early 1970s and, thus, involved new factors to which the global food system would eventually adjust. Additional causes of food insecurity external to the food system included the interrelation of the global financial crisis and food commodity prices. At one level, the dynamic of 'financialization' – where previously more tangible economic assets are converted into financial instruments and traded accordingly – is one of the signal characteristics of the world economy in the twenty-first century. In the case of 2008, however, the relative insecurity of traditional investment targets led to increased speculation on food commodity futures, exerting further upward pressure on prices. Such speculation represents the most extreme effect of the transformation of food into a commodity as its 'value' is determined on the basis of its potential return to investment relative to other commodities, with no recognition of its quality as an essential element of life. This situation, we argue, raises the essential issue of whether we treat food as a vital part of community, family and tradition, or as just another thing to be bought and sold.

In this book we also wish to suggest that the representation of the food crisis as an event, has undermined our ability to respond to global food security in positive and meaningful ways. To the extent that it assumes the capacity to maintain and reinforce a particular set of power relations in global society, the representation or descriptive narrative of a food crisis might be referred to as discourse. This discourse is then used as a rational basis to promote an inflexible, defined set of responses and policy solutions – in other words, the fact that a crisis occurred means that something is not working as it should. There has been a rupture. For the food crisis of 2008 the packaging of its causes within a descriptive phrase – the perfect storm – assumes that we know exactly what is not working. The result is a failure to acknowledge, let alone address, the injustices inherent to the existing food system. In the concluding chapter, we will argue that the concept of utopia – that is, envisioning an alternative food system founded in the recognition of the qualities of food, including the moral value of the right to food – offers a means to begin restructuring the food system in more just and flexible terms.

The argument that underlies the analysis in the following chapters is that, rather than an isolated event caused by a perfect storm of factors, the crisis was indicative of problems in the global food system as a whole. This is not a novel argument, but it is one that bears repeating with the essential need to shift away from business-as-usual. The approaches and positions utilized by the contributing authors to the book vary to some extent, but they share a common perspective such that an improved global food system requires radical changes to its organization. The following section introduces some of the theoretical background that contributes to the critical perspectives of the authors. An important element of these perspectives lies in identifying the crisis as a systemic effect that is likely to recur.

## The food crisis as point of departure

Food system critics conclude that business as usual cannot continue or, as Lang (2010, p97) argues, 'The crisis in 2005–8 was not a blip, but creeping normality'. The record food commodity prices and events in North Africa and the Middle East in early 2011 might bear this out. But just what does business as usual imply? To answer this question, we must briefly examine the historical development of the global food system.

To unravel the meteorology of the perfect storm we first have to understand how different our current food system is compared even to a mere 100 years ago. For the most part, countries and regions were food self-sufficient (or drew heavily on their colonial territories) and historic food crises related more to catastrophic environmental events – floods, droughts, disease – or despotic or colonial oppression, or a not so subtle combination of the two, as demonstrated by the Irish Potato Famine of the 1840s. The dramatic changes ushered in through the violent reorganization of geopolitics through two world wars and numerous other conflicts and the upending of historical colonialism led to massive structural changes in the food system following World War Two.

Since the late 1940s, the global food system has grown more complex (and volatile) in two major ways. First, countries in the Developed World shifted their policy towards food security and began a process of massive investment into agricultural intensification and industrialization within their domestic farm sectors. Second, corporations have consolidated the control of the system they began exercising as colonial plantations. At the same time, those corporations have grown larger, more complex and more powerful. Finally, the emergence of international food organizations (for example, the FAO) to oversee the intricate system implies a humanistic (read: *moral*) imperative to feed the world. Within the business-as-usual model, such organizations act as a moral foil to corporate consolidation. In other words, the corporations are free to profit from the production of food, leaving the governance structures to feed those that the markets bypass. Markets facilitate the movement of goods and ideas across vast areas of space. Prior to the emergence of market activity solely for profit (capitalism's defining feature), markets enabled the novel and the necessary to move like a gas – from constrained space into more open space. But the privileging of market reasoning for profit ignores the benefits and moral rationales for ensuring a nourished population. Even Adam Smith, he of the invisible hand, argued as much.

Within the business-as-usual model, food operates just like any other commodity. Related to the rhetoric of feeding the world, corporations or farmers-as-subcontractors are responsible for producing enough. And yet, the full picture is much murkier. The global food system has consolidated, putting more food under control of fewer firms. In the aftermath, the ideal of consumer choice erodes along with the potential for food self-provisioning.

In addition to the emergence of a powerful corporate sector in global food trading and the entrenchment of the neoliberal approach to global food relations during the last half century, other key actors and processes have contributed to the current crisis. For decades, governments and government-like organizations have negotiated, measured, assessed, identified and shifted the food landscape as they led us to believe that they were exterminating small fires of famine and hunger. The era of trust in government-led food solutions may well be over. The combination of the political and the technical solutions to feed the world yielded the green revolution and the subsequent hagiography of Norman Borlaug as a secular saint. A potentially insightful accounting exercise would involve calculating the global balance between the number of lives Borlaug 'saved' and the number of lives lost, resulting from diversion of food crops into cash crops, debt-driven poverty as farmers had to borrow to access more expensive technologies, farmer suicides in the Punjab, the lack of financial support into different kinds of agronomic research, vitamin-A deficiency and political instability. Absurd, right? But so is claiming that the green revolution saved so many people while the same number of – if not more – hungry people remain in the world.

The current state of the world food system therefore has many antecedents in the twentieth century: a particular trajectory of post-Second World War agricultural development in the Developed World; a steady move towards the centralization of market relations in the global economy; and governance mechanisms that sought to liberate the economy while solving those episodic problems that arose. All these created the historical conditions for a potentially catastrophic failure of the world food system. The global food events of 2008 may have demonstrated the turning of a global tide – the point at which the expansion of the market-based solution to global food provisioning reached its absolute limit and is now shrinking in the face of multiple ruptures at its now-decreasing margins.

This insight has not been absent from global debate, although its proponents have generally been somewhat disempowered. The unjustifiable pervasiveness of the hungry, the dispossessed and so many other 'externalities' within the business-as-usual model has engendered an 'opposing' side founded in smaller scales, agro-ecology, and food-related social movements. (We lay out this dichotomy here only as a means to get to the heart of the matter.) International groups like La Via Campesina and the Landless Peasants of Brazil (MST) articulate the position of farmers and rural communities in the Developing World. Both 'sides' agree on the goal of feeding the world. But the global movements for food sovereignty writ small, encompass social action to attain more access to international meetings, the assurance of fresh water, recognition of environmental limits and technical training for female farmers among others. This 'blessed unrest', as Paul Hawken (2007) terms it, is the largest social movement in the world.

It has not only been from the fringes that the current state of world food relations has been contested. Programmes such as the Millennium Development Goals and the UN Special Rapporteur for the Right to Food act as good faith initiatives to alleviate hunger, while their efforts are systematically undermined by the various relationships between states and corporations. The Declaration of Human Rights – the foundational document of the UN and an embodiment of the humanist project since the Enlightenment – has regrettably become mere window-dressing for the marketization of the world's food governance structures. In this case, critique is disempowered not at the margins, but at the very centre of global governance.

The neoliberal mantra – couched in the discourse of individual choice that underlies and justifies business-as-usual in the food system – argues that firms will facilitate the best distribution of resources and the greatest accumulation of wealth. And in one sense, neoliberalism is a great success. But at what cost? This is where the ideological gauntlet gets thrown down and we collectively avoid the moral discussion embedded in the assumptions on both sides.

Business-as-usual operates as an uneven playing field that is laden with doublespeak. On one hand, corporations and Developed World agricultural producers – US, UK, EU, Australia – promote free trade while subsidizing their farmers in various ways. In the US, subsidies for oil and ethanol skew the markets even more. At the other end, farmers and families are discouraged from the opportunity to grow appropriate varieties of staple crops for self-sufficiency in order to pursue the promises of the market and grow for a different audience. While Ricardo might relish the enactment of comparative advantage in the global food system, producers seduced by markets now face the volatility of oil prices, transportation costs, fickle consumer tastes and changing geopolitics. Here then is one of the moral dilemmas posed by the food crisis: how do we balance livelihood with the potential benefits of a roaring global economy?

## Food system critiques

Some food system analysts have used a political-economic approach (often informed by Marxist critiques of capitalism) to focus critique of business-as-usual on the power differentials it reinforces. From this perspective, the combination of disproportionate trade regulation (protectionism in practice, free trade advocacy for everyone else), the range of resource 'peaks' (oil, phosphorus, water, land) and global climate change provide a blueprint for difficult times if we continue to run a mode of agriculture based on infinite resources. Marx's theory of metabolic rift outlines the contradictions posed by capitalism's insatiable hunger for profit and accumulation and a planet composed of finite resources. The lack of regulation (national and international) to protect the resources of the commons (e.g. land, air, water) allows corporations, whose only constituency are shareholders, to literally horde, steal and plunder, while leaving a wake of tailings or bruised bananas in a community, without consequence. Corporate interests often contradict public interests as such. Countries employing protectionist trade policies in the food sector distort the market for food elsewhere.

According to political-economic analyses, the current global food system is sustained through an ideology that justifies exploitation of resources and labour. The underlying ideology of business-as-usual efforts to feed the world assumes that international food commodity markets operate on a level playing field. This implies that best operating procedures are the norm across the board such that commodity auctions are not subject to bribery, that trucks and roads will facilitate free passage of goods, customs officers will allow free thoroughfare for exchanged goods, that petroleum will be available, that seeds will be healthy and appropriate – these are assumptions on which most farmers in developed economies such as the US, UK, Canada, France and Australia can rely. But what about the farmers producing cassava in Peru, or sorghum in Uganda or export rice in Indonesia? And what about the large percentage of the world unable to grow, or unconcerned with growing, their own food? Consumers are just as victimized by uneven trading practices in global food. This contradiction between free market ideology and protectionist reality inhibits business-as-usual from delivering for everyone in the global food system.

The current global food system (including such aspects as food aid, famine relief, nutrition policy, etc.) generates uneven outcomes, manifest in vast wealth accumulation by corporate interests and massive hunger for over a sixth of the globe. Does such unevenness demonstrate market success? Or, pertinent to our purposes here, is much of that aid and the imperative to feed the world simply an 'externality cost' of deregulation – in essence, the pursuit of increasingly open markets and free trade is completely contrary to a world in which food security, not to mention food sovereignty, can be achieved.

Straightforward political-economic theories tend to underestimate the ability of firms and their relationships with governance structures to adapt. Enter regulation theory (originating in France) – or, as Bob Jessop argues, regularization theory. Regulation theory posits that capitalistic systems – firms, advocacy groups and their intersections with governments – manipulate legal systems and markets to balance the tension between insulating business and allowing for growth. The objective is to stabilize international markets to allow for accumulation, while also providing space for enough volatility to promote a competitive atmosphere for business. This is the ultimate tension between open markets and ensuring the common welfare.

Following the insights of regulation theory, it is apparent that any assumption of stability in the global food system is naive. The state of that system is subject to its position within the broader global economy. As the recent banking crisis reveals, profit seekers – mainly in the form of corporations, but represented by individuals – will push the limits of acceptable rent-seeking through financialization until told otherwise. In other words, the ability to make money depends on continually having new places to generate wealth – such as the creation of debt derivatives, which facilitate the frequent re-selling of the same debt so that every-body along the chain can accumulate capital, while the debt becomes worthless. Feeding the world with food commodities depends on people paying above cost for the products to ensure a profitable exercise. At the same time, when you push yourself over the precipice there's nowhere to go but down.

Harriett Friedmann and Phillip McMichael (1989) identified two such instances involving food where the system crumbled and then reorganized. They used the concept of food regimes to refer to the reorganized food systems that emerged in response to the collapse – a concept employed in several of the contributions to the first section of this book. The shift from the first to the second regime involved the implosion of imperial governance structures following the First World War. The second involved the geopolitical realignment following the Second World War. The question on the table asks whether the food crisis of 2008 (and 2011) represents a third major disruption?

While we will only be able to answer that question much later, our contributors all agree – food is different. It's not a car or a computer. As pointed out by Noah Zerbe (2009, p172): 'Perhaps food is unique among commodities in that it is fundamentally necessary for human experience. In this respect, any solution to the current crisis should begin with the premise that food should not be subject merely to regulation of the market; that food security should be a central goal of state policy' (p172). The adoption of this mindset – that sees food (and its provi-sioning) as a human right – engenders a host of different directions to reorient the global food system.

The contributors to this book join with a broad literature in arguing that – beyond the escalating commodity prices – the food crisis was also symptomatic of the social and environmental dimensions that have come to define the global food system. In other words, despite the strong desire to perceive the problem of hunger as solved, the existing means of producing food and distributing it to consumers involves deep-set challenges to social equity and environmental sustainability that are exposed by circumstances such as those playing out in the food crisis.

## Mapping the context of the food crisis

In the first section of the book, the contributors examine the broader contours and structures of the global food system. Each chapter employs a unique point of focus and draws on the approaches we have summarized above. Knitted together, these analyses establish features of the socio-ecological landscape which facili-tate food provision in some places and among some people, while conspiring to constrain availability of, and access to, food elsewhere. The common theme in these chapters is the underlying failure of the current global food system and the need to pursue alternative perspectives and objectives as the means of achieving global food security.

In the initial chapter of the first section, *Agriculture and food systems: our current challenge*, Jules Pretty draws on his extensive knowledge of food production systems, and experience with food governances to propose necessary solutions to the current failures of the global food system. The underlying premise for his recommendations is that the food system is highly complex, a factor further complicated by the socio-ecological nature of that system. Thus, progress to more appropriate means of producing the food to feed the global population must account for local contingencies as these relate to its environmental and social impacts. In other words, any solution to the current challenge of feeding the growing global population will involve more than a narrow focus on technologies for increasing the productive output of existing farming systems.

In the third chapter entitled *Let us eat cake?* Hugh Campbell employs the insights of food regime theory to examine the pernicious nature of food aid policies, which are ostensively oriented toward achieving global food security. He argues that the imperative of increasing the supply of food (that is, agricultural production) has been used to justify the subsidization of farmers in the developed world. While, in itself, promoting the economic viability of farming is not necessarily a problem, the implications of placing greater emphasis on domestic politics than global ethical responsibility are less sanguine. Campbell further demonstrates how this relationship to food aid enabled the justification of food as a geopolitical weapon, especially during the Cold War era.

While maintaining the food regimes approach, in contrast to the broad historical brush of Campbell's chapter, the contributions from Bill Pritchard and Philip McMichael focus on more specific moments of global food policy. In chapter 4 *Trading into hunger? Trading out of hunger? International food trade and the debate on food security*, Pritchard examines the contradictions between the pathways to achieving social good promoted in global trade, as compared to global food policy, respectively. His underlying argument is that the rules considered necessary to implement global free trade create legal and perceptual barriers to policies which might encourage food sovereignty, especially in the developing world. Chapter 5 by Philip McMichael *Biofuels and the financialization of the global food system* examines the issue of biofuels as the latest manifestation of the emphasis on commodities as the objective of agricultural production. He further examines the process of financialization of agricultural products, which he equates to a further abstraction of food from commodity to futures contract. The consequence of this change for global food security lies in the reduced recognition of food as an essential element of human life (its use value), thereby removing any connection to the ethical responsibilities associated with its exchange. Both of these chapters identify aspects of global food production that challenge local conditions of food security while also arguing for a more human rights-based emphasis in the economic regulation of the global food system.

Claire Mahon in chapter 6 *The right to food: a right for everyone* provides a shift to the strong sociological focus of the remaining chapters, arguing for the need to apply a human rights approach to the development of food policy. Based on her extensive experience as a member of Jean Ziegler's research group on the right to food, Mahon explains the implications of existing international human rights that include the right to food. It is in taking such obligations seriously that the world's governments can begin to bridge the policy contradictions identified by both Campbell and Pritchard.

In chapter 7, *Plentiful food? Nutritious food?* Colin Butler and Jane Dixon address a further, often unacknowledged, aspect of food security, namely the nutrition of the food available. They first examine the unequal access of societies and people to nutritious food and the emerging issue of overconsumption and obesity. This situation, they argue, is the result of a food production system that is poorly matched to the dietary behaviours and needs with which humans have evolved. They further locate the rationalization of the disjuncture identified by Raj Patel (2007) in *Stuffed and Starved* in Marx's concept of the metabolic rift, which suggests that consumers will become less aware of the social and environmental implications of production as they are more distanced from it.

The first section concludes with a more general argument about the misplaced (utopian) emphasis on quantity in the current food system. Paul Stock and Michael Carolan argue in chapter 8 *A utopian perspective on global food security* that the drive to feed the world has, to this point, focused on supply and demand as quantities in an equation that must be balanced. This emphasis ignores, however, the qualities inherent to food that necessitate greater awareness of processes of production and consumption. In this sense, Stock and Carolan provide further evidence of the type of shift in orientation and practice necessary to help account for the challenges to food security raised by the remaining authors in the first section.

## Real world experiences of the global food system

The second section of the book shifts the focus from cross-cutting themes involving the context of the food crisis, to case studies that address specific realities of the existing global food system. Each of these chapters examines a negative consequence of the food system as experienced in a specific location. They are also intended, however, to provide examples of the positive efforts of local food producers and consumers to overcome the constraints of the misplaced emphases and policies discussed in the first section.

In chapter 9 *Climate change and the resilience of commodity food production in Australia*, Geoff Lawrence, Carol Richards, Ian Gray and Naomi Hansar examine the impact of climate change on the Murray-Darling Basin in Australia. Both climate modelling and recent experience of extended drought suggest that a region once considered to be one of the world's breadbaskets will face serious climatic challenges to continued production. While farmers in the region will require ecologically appropriate production methods, Lawrence and his co-authors argue that such methods will not be seriously considered unless both the Australian Government and the agricultural industry acknowledge the need to temper productivism.

In comparison to the potential failure of an important source of exported food commodities in Australia, Jeff Neilson and Bustanul Arifin assess Indonesia's attempts to achieve self-sufficiency in domestic rice production. In chapter 10 *Food security and the de-agrarianization of the Indonesian economy*, they examine the economic costs of production subsidies that do not necessarily ensure accessibility or availability of food to the country's poor consumers. While acknowledging the logic of arguments for de-emphasizing self-sufficiency, Neilson and Arifin argue that the muted impact of the 2008 food crisis in Indonesia's domestic market suggests that such policy provides a valuable buffer in the face of increasing uncertainty in the global food system.

Chapter 11 by Nave Wald, Christopher Rosin and Doug Hill entitled *'Soyization' and food security in South America* focuses on commodity production of inputs to animal feed. They examine the impact of expanding areas of soybean cultivation in the Southern Cone of South America. While offering a means for governments to earn foreign exchange and wealthy land managers to realize high capital returns, soybeans also displace small-scale producers of diverse foods and cause severe environmental degradation. Wald, Rosin and Hill use the example of a peasant movement in northern Argentina to demonstrate the potential for cooperative response to the threats of the 'soyization' process and the implications for food sovereignty in the Southern Cone.

Kiah Smith and Kristen Lyons shift attention to small-scale producers in Africa in chapter 12 *Negotiating organic, fair and ethical trade: lessons from smallholders in Uganda and Kenya*. In this case, the focus is on two situations in which an emphasis on quality (organic and audit system defined products) offers a potential pathway to a more equitable engagement with the global food system. The quality focus of the products facilitates privileged access to European markets and promises more democratic production conditions. Smith and Lyons conclude that social issues of power inequalities related to gender and class remain to be overcome before the potential of this engagement can be realized.

Alec Thornton concludes the second section with an assessment of the development of urban agriculture in chapter 13 *Food for thought? Linking up urban agriculture and local food production for food security and development in the South Pacific*. In both Samoa and Fiji, tourism and industry have initiated a shift from customary food production systems on communal lands to market oriented production on the urban fringe. Thornton compares the systems of provision that have emerged around community-initiated farmers' markets and offers an insight to the potential for urban food production to contribute to domestic food security.

Each of the chapters in the book stands alone as a critique of a specific aspect of business-as-usual as it pertains to global food security. As a whole, they form a comprehensive condemnation of the global food system that has failed to remain true to the imperative to feed the world. We argue that the food crisis was more than a momentary blip, an event that developed as the result of a discrete set of new influences on the global market for food commodities. Something needs to be done – not merely to alleviate the vagaries of commodity prices, but also to re-establish the moral foundation of the imperative to feed the world. In the conclusion, we return to this point, arguing that this necessarily radical change involves a redefining of the utopia of a food-secure world. Such a utopian perspective is founded in the essentialization of food not as a commodity, but as a fundamental element of human existence.

## Note

1    The presentations and subsequent question and answer sessions from the 44th OFPS are available as podcasts entitled Dimensions of the Global Food Crisis through the University of Otago's iTunes U site (www.otago.ac.nz/news/itunesu/podcasts/ humanitieslectures.html).

## References

Cribb, J. (2010) *The Coming Famine: The Global Food Crisis and What We Can Do to Avoid It*, University of California Press, Berkeley, CA

IAASTD (2008) *International Assessment of Agricultural Knowledge, Science and Technology for Development*. Executive Summary of the Synthesis Report, Island Press, Washington, DC

Lang, T. (2010) 'Crisis? What Crisis? The Normality of the Current Food Crisis', *Journal of Agrarian Change*, vol 10, pp87–97

McMichael, P. (2007) *Development and Social Change: A Global Perspective*, Pine Forge Press, Newbury Park, CA

Patel, R. (2007) *Stuffed and Starved: The Hidden Battle for the World Food System*, Portobello Books, London

Zerbe, N. (2009) 'Setting the Global Dinner Table: Exploring the Limits of the Marketization of Food,' In J. Clapp and M. J. Cohen (eds.), *The Global Food Crisis: Governance Challenges and Opportunities,* Wilfred Laurier University Press, Waterloo, Canada, p161–175

# Part 1

# THE CONTRADICTIONS OF THE 'FEED THE WORLD' IDEOLOGY

# 2

# AGRICULTURE AND FOOD SYSTEMS: OUR CURRENT CHALLENGE

*Jules Pretty*

## Scale and immediacy of the challenge

Despite a significant growth in food production over the past half-century, one of the most important challenges facing society today is how to feed an expected population of some nine billion by the middle of the twentieth century. To meet expected demand for food without significant increases in prices, it has been estimated that we need to produce 70–100 per cent more food in light of the growing impacts of climate change and concerns over energy security (FAO, 2009; Godfray *et al.*, 2010). It will also require finding new ways to remedy inequalities in access to food. Today the world produces sufficient food to feed its population, but there remain more than one billion people who suffer from food insecurity and malnutrition (IAASTD, 2009).

This challenge is amplified further by increased purchasing power and dietary shifts in many parts of the globe, barriers to food access and distribution, particularly in the poorest regions, and pressure to meet the Millennium Development Goal of halving world poverty and hunger by 2015 (World Bank, 2007; Pretty, 2008; IAASTD, 2009; Royal Society, 2009). Despite the emergence of many innovations and technological advances in recent decades, this combination of drivers poses novel and complex challenges for global agriculture, which is under pressure to ensure food and energy security in ways that are environmentally and socially sustainable (National Research Council, 2010).

Complicating matters further, the past half-decade has seen growing volatility of food prices with severe impacts for the world's poor, most notably during the food price peaks of 2007–08 and 2010–11, and political and scientific controversy over the role that biofuels play (FAO, 2008; Searchinger *et al.*, 2008; Fargione *et al.*, 2008) in affecting carbon sinks and emissions. Indeed, land use change (for any purpose) is already implicated as a major driver of global change (Tilman *et al.*, 2001; Rockstrom *et al.*, 2009; Harvey and Pilgrim, 2010). Agricultural and food systems are estimated to account for one-third of global greenhouse gas emissions, more than twice that of the transport sector (IPCC, 2007). Thus the goal of the agricultural sector is no longer simply to maximize productivity, but to optimize across a far more complex landscape of production, rural development, environmental, and social justice outcomes (IAASTD, 2009; Godfray *et al.*, 2010; Sachs *et al.*, 2010).

The complexity of drivers facing global agriculture has received growing recognition (World Bank, 2007; Royal Society, 2009; National Research Council, 2010). However, there remain significant challenges to developing national and international policies that support the wider emergence of more sustainable forms of land use and efficient agricultural production, across both industrialized and developing countries (Pretty, 2008). A recent analysis of the top 100 global questions for agricultural and food systems (Pretty *et al.*, 2010) identified a series of interlocking and framing challenges for this century. They demonstrate the closely tied socioecological nature of agricultural and food systems, and show that solutions will have to come from more than one sphere of political, technological and economic life:

- climate change and water
- biodiversity and ecosystem services
- energy and resilience
- social capital and gender
- governance, power and policy making
- food supply chains
- consumption patterns.

### *Climate change and water*

Climate change predictions point to a warmer world within the next 50 years, yet the impact of rising temperatures on rainfall distribution patterns remains far less certain (IPCC, 2007). The situation for the oceans is equally serious, with coastal ocean temperatures documented to be warming 3–5 times more rapidly than the projections of the Intergovernmental Panel on Climate Change (IPCC), and the capacity of marine ecosystems to sequester one half of global carbon becoming impaired. From a global food security perspective, many commercial fish species are becoming economically extinct, with recent surveys showing 63 per cent of fish stocks globally needing intensive management toward rebuilding biomass and diversity due to exploitation (FAO, 2005). Interventions are required across the scale, from small fields to communities, watersheds, catchments and ultimately whole river basins, with a focus on increasing the productivity of both 'green' and 'blue' water use (Humphreys *et al.*, 2008). In some countries, 85 per cent of diverted water resources are now directed into agriculture with increasing competition for urban and industrial usage. For this reason, the need for improved crop, soil and water management practices, particularly in light of climate change, is growing.

## Biodiversity and ecosystem services

Agriculture has been a leading cause of loss of global biodiversity due to conversion of natural habitats, such as forests and wetlands, into farmland (Green *et al.*, 2005). Furthermore, the increased efficiency of agriculture has resulted in dramatic declines of many species using farmland habitats. Key drivers include the increased use of synthetic pesticides, herbicides and fertilizers, increased landscape homogeneity due to regional and farm-level specialization, drainage of waterlogged fields, loss of marginal and uncropped habitat patches, and reduction of fallow periods within arable rotations (Robinson and Sutherland, 2002; Wilson *et al.*, 2009). Moreover, the intensification of agriculture has been central to the degradation of ecosystem services, and has both increased the production of greenhouse gases and reduced levels of carbon sequestration (UNEP, 2010). The major challenge is to understand the best compromises between increasing food production while minimising the negative impacts on biodiversity, ecosystem services and society. Furthermore, new technologies will provide both a means of increasing the intensity of agriculture and the areas suitable for agriculture, for example through drought-resistant crops. Determining the best compromise requires an improved understanding of how to use new technologies, and the balance between intensification and extensification to ensure sustainable food production, ecosystem services, biodiversity and socio-economic impacts.

## Energy and resilience

As demand for energy grows in the coming decades, alternative energy sources will need to be identified to sustain the growing global population. Agriculture uses a considerable amount of energy, both directly in machinery and embedded within products used in agriculture (Schneider and Smith, 2009). The effects of high oil prices on low income rural households, and globally on agricultural inputs (pesticides and nitrogen fertilizers), transport, and tillage and irrigation systems, could produce declines in agricultural productivity, so exacerbating the pressures to expand the area of cultivated land at lower levels of productivity. Climate change will impact agriculture in many ways, some positive and some negative. The already significant challenge of producing more food using fewer inputs is exacerbated by the need for agriculture to adapt to climate change, while also reducing greenhouse gas emissions arising from agriculture in order to mitigate climate change (Smith and Olesen, 2010). Resilience to climate change will need to be a key property of sustainable agricultural systems in the coming decades, particularly in those regions projected to experience severe ecological shifts due to a changing climate.

## *Social capital and gender*

Social capital describes the importance of social relationships in cultural and economic life and includes such concepts as the trust and solidarity that exist between people who work in groups and networks, and the use of reciprocity and exchange to build relationships in order to achieve collective and mutually beneficial outcomes. Social capital is seen as an important pre-requisite to the adoption of sustainable behaviours and technologies over large areas, as well as a precondition for the sustainable management of certain resources and technologies. Farmer participation in technology development and participatory extension approaches have emerged as a response to such new thinking; farmer involvement enables novel technologies and practices to be learned directly and then adapted to particular agro-ecological, social and economic circumstances (Godfray *et al.*, 2010).

Changing agricultural research and development from the current bias toward male farmers to gender-equitability is not merely an issue of political correctness or ideology, it is a matter of development effectiveness that can benefit all of society. Creating gender-equitable agricultural research and development systems is a transformative intervention, leading to opportunities, commodities, relationships and services that ultimately change the way people do things. By understanding both the constraints and opportunities for women in agriculture, it will be possible to develop new ways to address their needs and enhance their contributions in order to improve agricultural productivity, food security, and poverty reduction (Meinzen-Dick *et al.*, 2010).

## *Governance, power and policy making*

Promoting agriculture for development presents a serious challenge of managing multiple agendas and the collective interests of formal and informal institutions (the state, the private sector, and civil society), together with their inter-relationships, obligations, processes, mechanisms and differences. It is precisely at this interface that governance, economic investment, power and policy-making converge and play critical roles. On governance, it is important to establish safeguards against risks, and assurances for the wellbeing and social and economic benefits of smallholders, where the state has an important role in influencing technology and policy options. Thus, external aid and delivery models, and the state's policy guidelines, are important issues (World Bank, 2007; Royal Society, 2009). To achieve rapid agricultural and rural growth requires a range of complementary investments across the broad spectrum of agricultural production systems, from the large mechanized, more intensive systems to smallholder units. Thus, questions about the best mix of public and private sector investments in irrigation and water management, rural roads, agricultural finance and extension services, among others, for the more intensive systems assume great importance (Lele *et al.*, 2010).

## Food supply chains

The food supply chain (FSC) encompasses all those activities that lie between on-farm production and the point of consumption. FSCs have experienced fundamental change since 1950, becoming increasingly global in extent and marked by upward trends in scale of production, number of lines of manufactured products and levels of economic concentration by sector. The governance of FSCs has, consequently, become more complex and multi-scalar, involving many public, private and civil society actors (Lang *et al.*, 2009). During the last two decades it has become increasingly apparent that the primary locus of power within the FSC has moved steadily downstream towards the buying desks of the major corporate food retailers (Vorley, 2003). Three-quarters of food sales in most industrialized countries now passes through supermarket checkouts. This has drawn critics to highlight the environmental implications of extended supply chains designed to achieve year-round provision at the lowest cost. Yet, this retail format is becoming increasingly prevalent worldwide, with rapid growth rates in many developing countries (Reardon and Gulati, 2008), and concerns raised about the dietary implications (Hawkes, 2008).

Work needs to be done to establish more precisely what 'sustainable food' represents. While life cycle assessment and other technical measures will be needed to evaluate energy, carbon and water footprints and other environmental impacts, social, economic and ethical criteria will also be required in calculating appropriate trade-offs (Edwards-Jones *et al.*, 2008; van Hauwermeiren *et al.*, 2007). Ultimately, the purpose is to better demonstrate the link between diet and environmental impact (Frey and Barrett, 2007) and social impact, thereby encouraging greater personal responsibility and behavioural change (Jackson *et al.*, 2008) in the development of more sustainable food supply chains.

## Consumption patterns

Increased purchasing power, shifting food preferences, access to global markets and growing populations have led to significant shifts in consumption patterns in recent years that are anticipated to continue into forthcoming decades. Daily per capita calorie consumption has increased from 2280kcal in the 1960s to 2800kcal shortly after the turn of the century. As income levels rise in developing countries, so it is expected that demand for meat will tend towards the per capita consumption rates of 115kg per year in the USA and 80kg per year in the UK (Royal Society, 2009). In China alone, meat consumption has more than doubled in the past 20 years, and is projected to double again by 2030 (Scherr and Sthapit, 2009). As a consequence of increasing demand, meat production is expected to grow from 229m tonnes in 1990 to 465m tonnes by 2050, and milk to grow from 580m to 1043m tonnes (Steinfeld *et al.*, 2006). Shifting consumption patterns combined with population growth have led to estimates that food production will be required to dramatically increase to meet growing consumption needs in the future (Lobley and Winter, 2010). The emergent pattern of dietary shifts is unlikely to provide the same health benefits as well-balanced diets rich in grains and other vegetable products. Increased meat and dairy consumption (particularly red meat), combined with increased intake of high sugar and high fat foods characteristic of modern, highly processed food products, are likely to lead to nutritional deficiencies as well as a growing number of cases of obesity and its associated illnesses, such as Type II diabetes and chronic heart conditions. This will increase demand for healthcare and require increased spending (Royal Society, 2009).

## The sustainable intensification of agriculture

It is clear that agricultural production gains across the world have helped millions of people to escape poverty, removed the threat of starvation, and provided a platform for rural and urban economic growth in many countries. Between 1961 and 2007, world agricultural production almost tripled whilst population doubled. The green revolution drove this production growth with new varieties, inputs, water management and rural infrastructure. Most increases in food production were achieved on the same agricultural land, with net area only growing by 11 per cent over this period.

All commentators now agree that food production worldwide will have to increase substantially in the coming years and decades (World Bank, 2007; IAASTD, 2009; Royal Society, 2009; UNEP, 2010; Godfray *et al.*, 2010). But there remain very different views about how this should best be achieved. Some still say agriculture will have to expand into new lands, but the competition for land from other human activities makes this an increasingly unlikely and costly solution, particularly if protecting biodiversity and the public goods provided by natural ecosystems (for example, carbon storage in rainforest) are given higher priority (MEA, 2005). Others say food production growth must come through redoubled efforts to repeat the approaches of the green revolution; or that agricultural systems should embrace only biotechnology or become solely organic. What is clear, despite these differences, is that more will need to be made of existing agricultural land. Agriculture will, in short, have to be intensified. Traditionally, agricultural intensification has been defined in three different ways: increasing yields per hectare, increasing cropping intensity (i.e. two or more crops per unit of land) or other inputs (water), and changing land-use from low-value crops or commodities to those that receive higher market prices.

It is now understood that agriculture can negatively affect the environment through overuse of natural resources as inputs, or through their use as a sink for waste and pollution. Such effects are called negative externalities because they impose costs which are not reflected in market prices (Baumol and Oates, 1988; Dobbs and Pretty, 2004). What has also become clear in recent years is that the apparent success of some modern agricultural systems has masked significant negative externalities, with environmental and health problems documented and recently costed for many countries (Pingali and Roger, 1995; Norse *et al.*, 2001;

Tegtmeier and Duffy, 2004; Pretty *et al.*, 2005). These environmental costs shift conclusions about which agricultural systems are the most efficient, and suggest that alternative practices and systems which reduce negative externalities should be sought. This is what Ken Giller has called the North-South divide between the 'effluents of affluence' and poverty caused by scarcity (Tittonell *et al.*, 2009)

Sustainable agricultural intensification is defined as producing more output from the same area of land while reducing the negative environmental impacts and at the same time increasing contributions to natural capital and the flow of environmental services (Pretty, 2008; Royal Society, 2009; Godfray *et al.*, 2010; Conway and Waage, 2010).

A sustainable production system would thus exhibit most or all of the following attributes:

- Utilizing crop varieties and livestock breeds with a high ratio of productivity to use of externally- and internally-derived inputs.
- Avoiding the unnecessary use of external inputs.
- Harnessing agro-ecological processes such as nutrient cycling, biological nitrogen fixation, allelopathy, predation and parasitism.
- Minimising use of technologies or practices that have adverse impacts on the environment and human health.
- Making productive use of human capital in the form of knowledge and capacity to adapt and innovate, and social capital to resolve common land-scape-scale problems.
- Quantifying and minimising the impacts of system management on externalities such as greenhouse gas emissions, clean water availability, carbon sequestration, biodiversity, and dispersal of pests, pathogens and weeds.

As both agricultural and environmental outcomes are pre-eminent under sustainable intensification, such sustainable agricultural systems cannot be defined by the acceptability of any particular technologies or practices (there are no blueprints). If a technology assists in efficient conversion of solar energy without adverse ecological consequences, then it is likely to contribute to the system's sustainability. Sustainable agricultural systems also contribute to the delivery and maintenance of a range of valued public goods, such as clean water, carbon sequestration, flood protection, groundwater recharge, and landscape amenity value. By definition, sustainable agricultural systems are less vulnerable to shocks and stresses. In terms of technologies, therefore, productive and sustainable agricultural systems make the best of both crop varieties, and livestock breeds and their agro-ecological and agronomic management.

The pioneering rice breeder, Peter Jennings (2007), who led early advancements in high yielding rice varieties during the first green revolution, has argued for an 'agronomic revolution': 'It is now widely recognized that rice yield gaps result from agronomic failings, and that future yield increases depend heavily on this science. Agronomy's time has come to lift farm productivity out of stagnancy'. Agronomy refers to the management of crops and livestock in their specific circumstances, and matches with the emergence of the term agro-ecology to indicate that there is a need to invest in science and practice that gives farmers a combination of the best possible seeds and breeds and their management in local ecological contexts.

This suggests that sustainable intensification will very often involve more complex mixes of domesticated plant and animal species and associated management techniques, requiring greater skills and knowledge by farmers. To increase production efficiently and sustainably, farmers need to understand under what conditions agricultural inputs (seeds, fertilizers and pesticides) can either complement or contradict biological processes and ecosystem services that inherently support agriculture (Royal Society, 2009; Settle and Hama Garba, 2011). In all cases farmers need to see for themselves that added complexity and increased efforts can result in substantial net benefits to productivity, but they need also to be assured that increasing production actually leads to increases in income. Too many successful efforts in raising production yields have ended in failure when farmers were unable to market the increased outputs. Understanding how to access rural credit, or how to develop warehouse receipt systems, and especially how to sell any increased output, becomes as important as learning how to maximize input efficiencies or build fertile soils. Equally, the creation of a social infrastructure of relations of trust, connections and norms is critical to effect and spread innovation.

## New forms of social infrastructure

Social capital is used as a term to describe the importance of social relationships in cultural and economic life. The term includes such concepts as the trust and solidarity that exists between people who work in groups and networks, and the use of reciprocity and exchange to build relationships in order to achieve collective and mutually beneficial outcomes. Norms of behaviour, coupled to sanctions, help to shape the behaviour of individuals, thereby encouraging collective action and co-operation for the common good.

The term 'social capital' captures the idea that social bonds and norms are important for people and communities. It emerged as a term after detailed analyses of the effects of social cohesion on regional incomes, civil society and life expectancy. As social capital lowers the transaction costs of working together, it facilitates cooperation. People have the confidence to invest in collective activities, knowing that others will also do so. They are also less likely to engage in unfettered private actions with negative outcomes, such as resource degradation. Collective resource management programmes that seek to build trust, develop new norms, and help form groups have become increasingly common, and such programmes are variously described by the terms community-, participatory-, joint-, decentralized-, and co-management.

Social capital is thus seen as an important prerequisite to the adoption of sustainable behaviours and technologies over large areas. Three types of social capital are commonly identified (Hall and Pretty, 2008). These are: the ability to work positively with those closest to us who share similar values, referred to as 'bonding social capital'; working effectively with those who have dissimilar values and goals is called 'bridging social capital'; finally, the ability to engage positively with those in authority either to influence their policies or garner resources, is termed 'linking social capital' (Woolcock, 1998; Pretty, 2003). Linking social capital encompasses the skills, confidence and relationships that farmers employ to create and sustain rewarding relationships with staff from government agencies. To gain the most from social capital, individuals and communities require a balanced mixture of bonding, bridging and linking relationships.

Where social capital is high in formalized groups, people have the confidence to invest in collective activities, knowing that others will do so too. Farmer participation in technology development, and participatory extension approaches have emerged as a response to such new thinking. New approaches such as Farmer Field Schools (FFS) and the Agricultural Knowledge and Information System (AKIS) have been developed, that emphasize the development of both social and human capital. As has been indicated above, direct farm level links between researchers, extensionists and farmers are a prerequisite for technology innovation and adaptation.

In the past, extension systems were seen as the means to link research outcomes to farmers. However, support for monolithic structures has declined, partly as a result of the limited success of transfer-of-technology styles of information flow (Anderson, 2008). Extension systems have been closed or underfunded, and thus many countries lack the institutions that can connect farmers with external agencies and markets. As a result, new forms of social infrastructure have emerged to build bonding, bridging and linking social capital. If trust between actors is good, then transformations in production systems are possible. New forms of farmer-based social infrastructure include Farmer Field Schools, cooperatives, Rural Resource Centres, business groups, common interest groups, micro-credit groups and catchment groups. Many of these help to build farmers' knowledge on particular areas, such as pests and diseases, or plucking intensity in tea. Farmers learn best when they are encouraged to experiment; researchers learn best when they work in a participatory way with farmers to ensure that plant materials and animals are suited to local needs and norms (e.g. through participatory plant breeding).

Farmer involvement in all stages of the innovation process is critical, as novel technologies and practices can be learned directly and then adapted to particular agro-ecological, social and economic circumstances. This is particularly important where a sustainable intensification practice or technology appears to break existing norms for farmers, such as introducing fodder shrubs into maize systems, grasses and legumes for pest management, early transplanting and wide-spacing of rice, and adoption of conservation agriculture that involves multiple innovations to replace ploughing. Farmer participatory research, on-farm testing, and farmer selection of plant materials have all been embedded in a number of institutions (Pretty *et al.*, 2011).

## Concluding comments

Rosin *et al.*, (this volume) show how different models of agricultural and food systems have emerged: aid regimes, US and European models, the New Zealand model, the green revolution and more. Yet the outcomes have been the same: large numbers of people are still hungry, environmental limits are being approached or have been passed, pressures on agricultural resources will grow, and consumers remain uncertain about how to exert their power and choice. The disturbing and remarkable annual obesity maps produced by the US Centers for Disease Control and Prevention show an alarming year-on-year increase in obesity in the US in just a single generation. Yet most adults know what to do - eat less, go out more. The challenge for the world food system centres on closing the gaps between what we know and what we must do differently. This means developing new and differing models, culturally-embedded and meaningful, that put food at the centre of economies and societies, and at the same time ensure that the growing and raising of food happens in ways that improve the environmental systems of the planet. Environmental limits are speaking: there is only the one planet. If people and their policy makers cannot be persuaded of this central and urgent proposition, then we are clearly in trouble. We have to produce much more food whilst doing no harm; we have to persuade richer people to eat less or at least eat sensibly; we have to find ways to eliminate the travesty of persistent and widespread hunger.

## References

Anderson, J. R. (2008) *Aligning Global Agricultural Research Investments with National Development Activities: The CGIAR Experience*, CGIAR Secretariat, Washington, DC

Baumol., W. J. and Oates, W. E. (1988) *The Theory of Environmental Policy*, Cambridge University Press, Cambridge

Conway, G. R. and Waage, J. (2010) *Science and Innovation for Development*, UKCDS, London

Dobbs, T. and Pretty, J. N. (2004) 'Agri-environmental stewardship schemes and "multi-functionality"', *Review of Agricultural Economics*, vol 26, no 2, pp220–237

Edwards-Jones, G., Plassmann, K., York, E. H., Hounsome, B., Jones, D. L. and Milá I Canals, L. (2008) 'Vulnerability of exporting nations to the development of a carbon label in the United Kingdom', *Environmental Science & Policy*, vol 12, pp479–490

FAO. (2005) *Review of the State of World Marine Fisheries Resources.* FAO Fisheries Technical Paper, No. 457, FAO, Rome

FAO (2008) *High-Level Conference on World Food Security: the Challenges of Climate Change and Bioenergy*, FAO, Rome. Available from www.fao.org/foodclimate/hlc-home/en/

FAO (2009) *World Summit on Food Security*, FAO, Rome. Available from www.fao.org/wsfs/world-summit/en/

Fargione, J., Hill, J., Tilman, D., Polasky, S. and Hawthorne, P. (2008) 'Land clearing and the biofuel carbon debt', *Science*, vol 319, pp1235–1238

Frey, S. and Barrett, J. (2007) 'Our health, our environment: The ecological footprint of what we eat', *Paper Prepared for the International Ecological Footprint Conference*, Cardiff, 8–10 May 2007

Godfray, C., Beddington, J. R., Crute, I. R., Haddad, L., Lawrence, D., Muir, J. F., Pretty, J., Robinson, S., Thomas, S. M. and Toulmin, C. (2010) 'Food security: the challenge of feeding 9 billion people', *Science*, vol 327, pp812–818

Green, R. E., Cornell, S. J., Scharlemann, J. P. W. and Balmford, A. (2005) 'Farming and the fate of wild nature', *Science*, vol 307, pp550–555

Hall, J. and Pretty, J. (2008) 'Then and now: Norfolk farmers' changing relationships and linkages with government agencies during transformations in land management', *Journal of Farm Management*, vol 13, no 6, pp393–418

Harvey, M. and Pilgrim, S. (2010) 'Competition for land: food and energy', *Paper Prepared for UK Government Foresight Project on Global Food and Farming Futures*, August 2010. UK Government, London

Hawkes, C. (2008) 'Dietary implications of supermarket development: A global perspective', *Development Policy Review*, vol 26, no 6, pp657–692

Humphreys, E., Peden, D., Twomlow, S., Rockström, J., Oweis, T., Huber-Lee, A. and Harrington, L. (2008) *Improving Rainwater Productivity: Topic 1 Synthesis Paper*, CGIAR Challenge Program on Water and Food, Colombo

IAASTD. (2009) *Agriculture at a Crossroads*, International Assessment of Agricultural Knowledge, Science and Technology for Development, Island Press, Washington, DC

IPCC (Intergovernmental Panel on Climate Change). (2007) 'Chapter 11. Regional Climate', In Solomon, S., Quin, D., Manning, M., Chen, Z., Marquis, M., Averyt, K. B., Tignor, M. and Miller, H. L. (eds), *Climate Change 2007: The Physical Science Basis*, Contribution of Working Group 1 to the Fourth Assessment Report of the Intergovernmental Panel on Climate Change, Projections, Cambridge University Press, Cambridge

Jackson, P., Ward, N. and Russell, P. (2008) 'Moral economies of food and geographies of responsibility', *Transactions Institute of British Geographers*, vol 34, pp12–24

Jennings, P. (2007) 'Luck is the residue of design', *The IRRI Pioneer Interviews*, IRRI, Philippines

Lang, T., Barling, D. and Caraher, M. (2009) *Food Policy: Integrating Health, Environment and Society*, Oxford University Press, Oxford

Lele, U., Pretty, J., Terry, E. and Trigo, E. (2010) *Transforming Agricultural Research for Development*, Global Author Team, The Global Forum for Agricultural Research (GFAR), Report for the Global Conference on Agricultural Research (GCARD), 28–31 March 2010, Montpellier, France

Lobley, M. and Winter, M. (eds) (2010) *What is Land for? The Food, Fuel and Climate Change Debate*, Earthscan, London

MEA (Millennium Ecosystem Assessment) (2005) *Current state and trends*, Washington, DC

Meinzen-Dick, R., Quisumbing, A., Behrman, J., Biermayr-Jenzano. P., Wilde, V., Noordeloos, M., Ragasa, C. and Beintema, N. (2010) 'Engendering Agricultural Research', *Paper prepared for Global Conference on Agriculture and Rural Development* (GCARD), 28–31 March, 2010, Montpellier, France

National Research Council (2010) *Toward Sustainable Agricultural Systems in the 21st Century*, National Research Council Report, The National Academies Press, Washington, DC

Norse, D., Ji, L., Leshan, J. and Zheng, Z. (2001) *Environmental Costs of Rice Production in China*, Aileen Press, Bethesda

Pingali, P. L. and Roger, P. A. (1995) *Impact of Pesticides on Farmers' Health and the Rice Environment*, Kluwer, Dordrecht

Pretty, J. (2003) 'Social capital and the collective management of resources', *Science*, vol 302, pp1912–1915

Pretty, J. (2008) 'Agricultural sustainability: concepts, principles and evidence', *Philosophical Transactions of the Royal Society of London B*, vol 363, no 1491, pp447–466

Pretty, J., Lang, T., Ball, A. and Morison, J. (2005) 'Farm costs and food miles: an assessment of the full cost of the weekly food basket', *Food Policy*, vol 30, pp1–20

Pretty, J., Sutherland, W. J., Ashby, J., Auburn, J., Baulcombe, D., Bell, M., Bentley, J., Bickersteth, S., Brown, K., Burke, J., Campbell, H., Chen, K., Crowley, E., Crute, I., Dobbelaere, D., Edwards-Jones, G., Funes-Monzote, F., Godfray, H. C. J., Griffon, M., Gypmantisiri, P., Haddad, L., Halavatau, S., Herren, H., Holderness, M., Izac, A.-M., Jones, M., Koohafkan, P., Lal, R., Lang, T., McNeely, J., Mueller, A., Nisbett, N., Noble, A., Pingali, P., Pinto, Y., Rabbinge, R., Ravindranath, N. H., Rola, A., Roling, N., Sage, C., Settle, W., Sha, J. M., Shiming, L., Simons, T., Smith, P., Strzepeck, K., Swaine, H., Terry, E., Tomich, T. P., Toulmin, C., Trigo, E., Twomlow, S., Vis, J. K., Wilson, J. and Pilgrim, S. (2010) 'The top 100 questions of importance to the future of global agriculture'. *International Journal of Agricultural Sustainability*, vol 8, pp219–236

Pretty, J., Toulmin, C. and Williams, S. (2011) 'Sustainable intensification in African agriculture', *International Journal of Agricultural Sustainability*, vol 9, pp5–24

Reardon, T. and Gulati, A. (2008) *The supermarket revolution in developing countries: Policies for 'competitiveness and inclusiveness'*, IFPRI Policy Brief 2, June. International Food Policy Research Institute, Washington, DC

Robinson, R. A. and Sutherland, W. J. (2002) 'Changes in arable farming and biodiversity in Great Britain', *Journal of Applied Ecology*, vol 39, pp157–176

Rockstrom, J., Steffen, W., Noone, K., Persson, A., Chapin, F. S. III, Lambin, E. F., Lenton, T. M., Scheffer, M., Folke, C., Schellnhuber, H. J., Nykvist, B., de Wit, C. A., Hughes, T., van der Leeuw, S., Rodhe, H., Sörlin, S., Snyder, P. K., Costanza, R., Svedin, U., Falkenmark, M., Karlberg, L., Corell, R. W., Fabry, V. J., Hansen, J., Walker, B., Liverman, D., Richardson, K., Crutzen, P. and Foley, J. A. (2009) 'A safe operating space for humanity', *Nature*, vol 461, pp472–475

Royal Society (2009) *Reaping the Benefits: Science and the Sustainable Intensification of Global Agriculture*, RS Policy Document 11/09, The Royal Society, London

Sachs, J. D., Remans, R., Smukler, S., Winowiecki, L., Andelman, S. J., Cassman, K. G., Castle, D., DeFries, R., Denning, G., Fanzo, J., Jackson, L. E., Leemans, R., Lehmann, J., Milder, J. C., Naeem, S., Nziguheba, G., Palm, C. A., Pingali, P. L., Reganold, J. P., Richter, D. D., Scherr, S. J., Sircely, J., Sullivan, C., Tomich, T. P. and Sanchez, P. A. (2010) 'Monitoring the world's agriculture', *Nature*, vol 466, pp558–560

Scherr, S. J. and Sthapit, S. (2009) 'Farming and land use to cool the planet', *State of the World 2009: Into a Warming World*, Worldwatch Institute, Washington, DC

Schneider, U. and Smith, P. (2009) 'Energy intensities and greenhouse gas emission mitigation in global agriculture', *Energy Efficiency*, vol 2, pp195–206

Searchinger, T., Heimlich, R,. Houghton, R. A., Dong, F., Elobeid, A., Fabiosa, J., Tokgoz, S., Hayes, D. and Yu, T. (2008) 'Use of US croplands for biofuels increases greenhouse gases through emissions from land-use change', *Science*, vol 319, pp1238–1240

Settle, W. and Hama Garba, M. (2011) 'The FAO Integrated Production and Pest Management Programme in the Senegal and Niger River Basins of Francophone West Africa', *International Journal of Agricultural Sustainability*, vol 9, pp171–185

Smith, P. and Olesen, J. E. (2010) 'Synergies between mitigation of, and adaptation to, climate change in agriculture', *Journal of Agricultural Science*, vol 148, pp543–552

Steinfeld, H., Gerber, P., Wassenaar, T., Castel, V., Rosales, M. and de Haan, C. (2006) *Livestock's Long Shadow: Environmental Issues and Options*, FAO, Rome

Tegtmeier, E. M. and Duffy, M. D. (2004) 'External costs of agricultural production in the United States', *International Journal of Agricultural Sustainability*, vol 2, pp1–20

Tilman, D., Fargione, J., Wolff, B., D'Antonio, C., Dobson, A., Howarth, R., Schindler, D., Schlesinger, W. H., Simberloff, D. and Swackhamer, D. (2001) 'Forecasting agriculturally driven global environmental change', *Science*, vol 292, no 5515, p281

Tittonell, P., van Wijk, M. T., Herrero, M., Rufino, M. C., de Ridder, N. and Giller, K. E. (2009) 'Beyond resource constraints – exploring the biophysical feasibility of options for the intensification of smallholder crop-livestock systems in Vihiga district, Kenya', *Agricultural Systems*, vol 101, pp1–19

UNEP (2010) *Assessing the Environmental Impacts of Consumption and Production: Priority Products and Materials*, A Report of the Working Group on the Environmental Impacts of Products and Materials to the International Panel for Sustainable Resource Management, UNEP, Nairobi

van Hauwermeiren, A., Coene, H., Engelen, G. and Mathijs, E. (2007) 'Energy lifecycle inputs in food systems: A comparison of local versus mainstream cases', *Journal of Environmental Policy and Planning*, vol 9, pp31–51

Vorley, B. (2003) *Food, Inc.: Corporate Concentration from Farm to Consumer*, UK Food Group, IIED, London

Wilson, J. D., Evans, A. D. and Grice, P. V. (2009) *Bird Conservation and Agriculture*, Cambridge University Press, Cambridge

Woolcock, M. (1998) 'Social capital and economic development: towards a theoretical synthesis and policy framework', *Theory and Society*, vol 27, pp151–208

World Bank (2007) *World Development Report 2008: Agriculture for Development*, World Bank, Washington, DC

# 3

# LET US EAT CAKE? HISTORICALLY REFRAMING THE PROBLEM OF WORLD HUNGER AND ITS PURPORTED SOLUTIONS

*Hugh Campbell*

## Introduction

In this chapter I address one of the core deceits of post-Second World War agricultural policy in the Developed World: namely, that seeking solutions to world hunger would be a central focus of international political efforts in the post-Second World War period. I argue for an alternative understanding of these political efforts: that is, that the central political concern in Developed World agricultural policy has not been to solve world hunger, but to solve the problem of how to secure Developed World incomes for Developed World farmers. While there is considerable rhetoric mobilized to suggest that these two goals are one and the same, this chapter will use the theoretical device of Food Regimes to explore some key dynamics in the longer term history of global agriculture that demonstrate that this is anything but the case. Starting with examination of the dynamic power of the 'food scarcity' problem of the industrial nations in the nineteenth century, this chapter will introduce the century-long evolution of a global solution to food scarcity in the core industrializing countries in what would become known as the Developed World. It will then go on to demonstrate how this long pattern of stable food arrangements was completely overturned in the period after the Second World War, resulting in three historically and geographically contingent agricultural, trade and wider economic policy regimes in the Developed World; regimes that have attempted to seek a solution to how to secure incomes for Developed World farmers. These are: 1) the post-Second World War aid-based regime encouraging the expansion of the US model of export-commodity platforms within the US and abroad; 2) the European model of direct subsidization of farmers, first through direct price supports, and then through 'green box' subsidies providing indirect income to farmers to support 'multifunctionality'; and 3) the neoliberal model demonstrated in countries like New Zealand, where unsubsidized, export agriculture has been restructured in an attempt to reposition Developed World production towards high-value, elite, quality-defined markets for agricultural products. At least two of these models use the rhetoric of 'solving world hunger' as a justification, but all are, on closer examination, actually specifically directed towards supporting or improving farm incomes in the Developed World. Solving world hunger remains, at best, a side effect of the central political intent of these models of Developed World agriculture. For this reason, all three models fail to provide a template for solving world hunger on a global scale.

## The Industrial Revolution and food scarcity

At the outset of the Industrial Revolution – somewhere in the latter part of the eighteenth century – food scarcity, famine, hunger and death by starvation were features of all societies and cultures around the world. Climatic variability, the unreliable quality of some human patterns of food production, war, unrest, colonization, slavery and land degradation all contributed to the relatively frequent appearance of times of famine and hunger in all parts of the world. The ever-present threat of death by starvation was one of the unifying conditions of humankind. The Industrial Revolution signalled a dramatic transformation of this universal reality, eventually banishing food scarcity from the industrial world (even if food availability became highly uneven across classes) and rendering hunger a problem that was manifest 'out there', at a safe distance from industrializing societies.

This transition was not merely a side-effect of the fundamental changes in social structure that were integral to the Industrial Revolution. Rather, food and global-scale food relationships were at the heart of the Industrial Revolution. The celebrated anthropologist Syd Mintz argues in his greatest work *Sweetness and Power* (1985) that food is the neglected element in many theories explaining the scale and trajectories of the Industrial Revolution in the eighteenth and nineteenth centuries. Using sugar as his central theme, Mintz outlines how the Industrial Revolution drew in global-scale webs of food relationships around commodities like sugar. These webs linked devastated tribal societies in Africa to slave plantations in the Caribbean, to rum producers and fisherman in New England, through to the emerging working classes in industrial England (see also Wolf, 1982). Central to this nexus of dynamism and devastation was a central and compelling concern for the ruling classes of Britain – how to feed the massively expanding working class population that provided cheap labour for the Industrial Revolution.[1] Britain was in the midst of one of the most tumultuous transformations in human history: one part of that dynamic was the rapid shift in population from being almost entirely rural (and living close to food sources) to being almost entirely urban (and separated from easily available food). This transition was by no means easy, and the 1840s became known as the 'Hungry Forties' as an exploding urban population outstripped the capacity of British farmers to produce food (even in a period of massive expansion in agricultural productivity), leading to bread riots in the cities of the English Midlands and a progressive stripping of all transportable food crops from nearby colonies like Ireland (leaving the local populace dependent on the potato – with devastating effects in the potato famine of 1845/46 (Salaman, 1949)).

The Hungry Forties emerge as one of the most pivotal decades in defining the long term trajectories and impacts of the Industrial Revolution on global food and agricultural systems. Central to this effect was the repeal of the Corn Laws in 1846, purportedly to 'assist' with the Irish Potato Famine, but actually signifying a much deeper shift in the political intent of colonial Britain. At face value, the Corn Laws had been a bastion of the entrenched economic privileges of the landed gentry and new capitalist farming class who received a guaranteed income protected by border tariffs on imported grains. The ability of the British Parliament to repeal the Corn Laws in 1846 signalled not only a shift in economic power from rural to urban Britain, but also demonstrated the power of the 'French example'. In the lifetimes of most parliamentarians, memories were strong in terms of what could happen in even the most wealthy European powers if a large urban populace was insufficiently well fed. Rather than reproduce Marie Antoinette's less than successful strategy to 'Let them eat cake', the British Parliament proposed that the solution to bread riots in Manchester was to 'Let them eat imported grain'. It was best to conveniently ignore that this grain was being imported from the Irish colony which was, at that very moment, being wracked by one of Europe's most harrowing famines.

A subsequent revolution in global transport infrastructure (from canals to railways, steamships, and motorized transport) opened up a vast range of possibilities for sourcing imported grain. In 1863, Anthony Trollope wrote of grain flowing in 'rivers' out of the American Mid-West to feed the Industrial Revolution. At the same time, this river of grain was joined by other significant tributaries running from the Canadian prairies, Argentine pampas, Australian river valleys, New Zealand grasslands, Indian floodplains and the Russian steppes. By the 50th anniversary of the repeal of the Corn Laws, the map of global food production had been transformed with the creation, incorporation and mobilization of vast empires in the service of providing the means to avoid food scarcity in the expanding industrial/imperial core.

Mike Davis argues in *Late Victorian Holocausts* (2001) that the true brutality of this reconfiguration of global food production was only vaguely visible from the core industrial empires driving this process. The elimination of hunger within Victorian society in Britain provided a legitimate moral justification for ignoring tens of millions of deaths by famine in the distant edges of the Empire. The success of that mission was such that, by 1900, Britain was supporting an urban population some six times larger than that of 1800; while over the subsequent decades, structural hunger and food deficits were only infrequent visitors.

In summary, the century that stretched from the Hungry Forties to the Second World War saw the establishment and entrenchment of a pattern of global food production, trade and consumption that effectively ended food scarcity as a structural feature of societies in the Developed World. At the same time, on the periphery, hunger and famine continued (and in Mike Davis's analysis became considerably more entrenched) as a feature of life. The previously universal experience of hunger and food scarcity was no longer, it seemed, something common to all humankind.

This century of stable configuration of world food relationships did not endure. What happened next was the subject of a highly influential article by Harriet Friedmann and Philip McMichael. Writing in 1989, these two authors sought explanations as to how (and why), after a long period of stable configuration of global political and agri-food arrangements, food relationships, global trade patterns and agricultural policy were completely reconfigured (Friedmann and McMichael, 1989). Their idea of transition between historically configured 'food regimes' highlighted a dramatic transformation of global food arrangements in the period between the start of the Great Depression and the immediate post-Second World War years. This transformation was partly caused by dramatic shifts in economic policy and goals during the Great Depression, and partly by the traumatising effect of food scarcity and rationing in Britain during the mid-years of the war as German U-boats effectively cut Britain off almost completely from its colonial food supply.

The agricultural policy shift in the Developed World that emerged during this period – accompanied by multiple transformative shifts in political orientation, economic goals and foreign policy during the early decades of the Cold War – established the conditions under which the main focus of these shifts can be examined: the need to encourage and secure the livelihoods of farmers in the Developed Countries themselves.

## Towards food security: the aid regime, PL480 and feeding the (non-communist) world

Friedmann and McMichael's (1989) key insight, which effectively transformed the way in which the global political economy of agriculture was being theorized at the time, was that agriculture and capitalism were not emerging in some kind of linear historical process. This idea of linear progress mirrored a similar narrative in environmental politics that attributed incremental, ongoing and deleterious impacts on the environment to a process called 'industrialization' that commenced with the Industrial Revolution and has intensified inexorably ever since. Friedmann and McMichael (1989) posited that there had actually been a number of quite different configurations of the global agricultural regime with periods of crisis and dramatic reconfiguration identifiable over the longer term. Put simply, the history of agriculture under industrialization was not just 'one damn thing after another' – actually, it was lurching between states of stability and periods of crisis.

This chapter has already identified a particular set of food relations and conditions that emerged around the Industrial Revolution and eventually incorporated food producers at a global scale. Friedmann and McMichael (1989) describe this set of relations as the 'First Food Regime' (later describing it as the 'Imperial

Food Regime'). One of the claimed achievements of each regime was that internal contradictions and tensions were managed or contained (or ignored) in ways that enabled the long term stability of the regime. Eventually, these contradictions start to become destabilizing and overwhelming, and Friedmann and McMichael (1989) argue that the First Food Regime moved into its final decline and crisis during the Great Depression and the Second World War. Effectively, these two global-scale crises brought to an end a long trajectory of agricultural (and wider trade policy) development that began with the repeal of the Corn Laws, rose through the free trade era of imperial expansion in the nineteenth century, and embedded a global model of food relations in which wealthy industrial countries fed off their colonies around the world.

This moment of transition and crisis shifted world food relations to what Friedmann and McMichael (1989) termed the 'Second Food Regime'. In this shift, a fundamental change occurred in the political significance of actual agricultural producers in the core countries themselves. What had become apparent during the Great Depression was that, after a century of focus on policy regimes to support industrialization, core industrial countries like the USA still actually relied on some level of economic prosperity in the domestic agriculture sector as well. Attempts to reconstruct Europe (poorly) after the First World War were demonstrating a similar vulnerability of industrial economies to weakness or stagnation in domestic agriculture sectors. Finally, the various traumas of the Second World War, particularly the effective isolation of Britain from its primary food supply (abroad) by the German U-Boat fleet, and the massive political intervention required to prevent some in Britain from starving, all contributed to a major shift in political thinking around the importance of domestic agriculture sectors in industrial countries. The emerging threat of the Cold War strengthened this resolve, given that the potential of the Soviet bloc to isolate Western Europe from its global food supply was even more pronounced than the blockade of Britain in the Second World War. In the 1950s, during that period prior to massive proliferation of nuclear weapons which made such scenarios less appealing, Soviet military strategists even calculated the putative number of days that Western Europe could sustain itself on domestic food sources during a potential continental blockade.

The answer to this challenge was to dramatically shift focus back onto domestic agriculture and food production in the industrial core countries. Friedmann and McMichael (1989) identify the key political shift post-Second World War as being towards supporting food security by encouraging domestic agriculture. In industrial countries this meant a major reconfiguration of agricultural policy. Beginning with a cluster of legislation around the UK Agricultural Act of 1947, and coupled with the US-led Marshall Plan in Europe, policy rapidly evolved to direct state subsidization towards agricultural production at home. Government subsidies to increase the rate of agricultural mechanization; incentives to use fertilizers; adopt new breeds and varieties; consolidate landholdings; and undertake skills development in the emerging techniques of industrial/intensive agriculture were partnered with state investment and support to enable the transition of wartime industry into agricultural applications (explosives into fertilizers, tanks into tractors, gases into pesticides).

By the latter stages of the 1950s (in the US) and the mid-1960s (in UK/Europe), these policies were demonstrably successful. During the 1950s, the Marshall Plan ceased to act as the conduit for agricultural surpluses from the US into the recovering economies of Europe. Consequently, a problem of surplus production began to arise. US agriculture under industrialization was, it seemed, just too productive. At some point in the 1960s, a similar dynamic emerged in the UK/Europe as agricultural productivity surged and subsidized production of key commodities moved into surplus. From this point the phenomenon of 'grain stockpiles', 'butter mountains' and even 'wine lakes' became a problem for European politicians to confront. In both the US and the European case, however, there was no sense that things had overshot the mark. Rather, this new, highly productive domestic agriculture sector was seen as creating all sorts of new strategic and political opportunities.

This new dynamic of surplus food production on both sides of the Atlantic created what Friedmann and McMichael (1989) initially termed the 'Second Food Regime', but later also termed the 'Cold War' or 'Aid-Based Food Regime'. Several authors, including Susan George in *How the Other Half Dies* (1976), identify the US legislation PL480 (later termed the 'Food for Peace' law) and its various subsequent revisions as symbolically demonstrating this major shift in world agricultural relations. Through PL480, the US (with parallel mechanisms later implemented by the UK/Europe) began to use the instrument of food aid as an element of foreign policy. Surplus food production in the Developed Countries was not only sold to the Developing World, but it also began to be distributed widely as aid. Food aid became a key element in the development project for these societies through enabling the rural peasantry (who were no longer required to produce quite so much food) to migrate to new urban mega-cities (where they could be fed by subsidized grain imports and gifts from places like the USA). Within a few years of the passing of PL480, around a quarter of all US grain exports were taking place under the auspices of such aid programmes.

Needless to say, this dramatic new initiative became part of a wider set of Cold War strategies to win (or coerce) the support of Developing World countries – many now ex-colonies – for either side. Whereas in the early 1950s, strategists had calculated the number of days of available food supply in Western Europe, within a decade the key strategy had shifted towards the most creative use of food surpluses. As George (1976) documents, a suspiciously large number of the key recipients of PL480 aid were mainly notable for being Western allies during the Cold War (like Pakistan) rather than for actually featuring on the UN list of food deficit countries.

One essential element within the Second Food Regime which Friedmann and McMichael (1989) and Friedmann (1990, 1993) identified also supports a key contention of this chapter: that securing the livelihoods of Developed World agriculturalists shifted from being almost irrelevant in the First Food Regime to being a central plank of state policy and investment during the Cold War. It became absolutely imperative to ensure food security in the Developed World, with the resulting surpluses being seen as a global strategic opportunity rather than an economic problem. In this context, what is clearly revealed is that the subsequent 'feed the world' rhetoric that animated the Western mission to the Developing World during the Cold War was inextricably linked, and partly originated from, the need to find a way to secure the incomes and productivity of Developed World farmers themselves. This need stood prior to, and superior to, the need to feed the hungry of the world. Put simply, faced with a mounting problem of world hunger post-Second World War, the response was not so much 'let them eat cake' as 'let them eat our cake': this resulted in the technocratic and policy challenge of how to ensure that Developed World farmers were able to produce lots of cake!

## Three emerging models for securing Developed World incomes for Developed World farmers

The answer to the problem of securing livelihoods for Developed World farmers took three forms – all reflecting the different positioning of key countries in the Developed World during the Cold War years. While the specific history of agri-culturally-related trade negotiations is outlined in the chapter by Pritchard (this book), the key point for this argument is that the necessity of accommodating varying policy trajectories and models had been accepted by the Western powers post-Second World War, and meant that agriculture was actually exempted on food security grounds from being discussed under any of the GATT (General Agreement on Tariffs and Trade) negotiations until the Uruguay Round in 1985.

As outlined in the previous section, the first decisive move into the policy mechanisms that would underpin the Second Food Regime emerged in the US.

### The US model

The US had, since the earliest years of industrialization, retained a particular economic configuration that reflected the fact that a substantial portion of the US was geographically well-suited to sustain farming activities. Hence, during the New Deal, specific measures were directed at supporting agricultural incomes. During the Second World War and in the years of the Marshall Plan, US agriculture was subsidized through three mechanisms instituted around the Great Depression: the Grain Futures Act (1922); the Agricultural Marketing Act (1929); and the major Agricultural Adjustment Act (1933). A main focus of these Acts (reflected in subsequent legislation) was to secure the production of major commodities through both direct price support and export subsidies. Secondary subsidy mecha-nisms that were deployed included periods of 'set-aside' and the creation of new demand for export commodities through aid (although this was prefigured in the National School Lunch Act of 1946).

While US agricultural policy is complex, and subject to constant subtle changes due to the power of agricultural lobbies and the pivotal nature of farming constituencies, the general US model for securing the livelihoods of farmers has been to encourage the establishment and maintenance of major commodity platforms, with the primary direction for those products being overseas markets. The ability of US farmers to access continually expanding destination markets for their products was enabled not only by aggressive trade policy in formal markets, but also by the strategic deployment of aid initiatives to gain access to consumers outside the reach of the formal marketplace.

This formation was not only created as a result of policy initiatives. Post-Second World War conditions were also highly advantageous for corporate investment into agriculture – with guaranteed prices, massive international demand (particularly during the Marshall Plan) and significant interventions to ensure the maximum possible extent of market scope and access (including through food aid). The result was a massive movement of corporate investment into US agriculture at that time, with major agri-corporations emerging around both agricultural input industries (like Dow and Monsanto) and the processing and trading of the major grain commodities (like ConAgra and ADM). These new agri-corporates became key contributors to the expansion and elaboration of the wider Second Food Regime. Many consequences flowed from this propitious alignment of policy and corporate interests. First, these corporates (or their forebears) were vociferous supporters of the green revolution in the 1960s and 70s, as they stood to be the immediate beneficiaries of the new markets opening up for agricultural technologies in the Developing World, as well as having the opportunity to trade in the resulting commodities on global markets. While the green revolution was widely considered a failure in terms of solving the problem of world hunger, it was, nevertheless, a significant success in ensuring the profitability of the high-tech corporate model of agriculture – and its farmers. It has also been credited with entrenching a particular model of intensive industrial agriculture that is ecologically unsustainable (McNeill, 2000).

The economic, institutional, and political power of these corporate entities (as well as the wider farming constituencies that became embedded around this particular model of Second Food Regime agriculture) has ensured that the subsidized, export/commodity approach to sustaining US farmer livelihoods has endured well beyond the Cold War – and through a series of crises, wider trade policy shifts, political administrations and changes of economic orthodoxy. It largely endures not because it helps solve world hunger in its destination markets, but because it secures a particular level of livelihood for US farmers and a particular level of profitability for US agri-corporates.

### The European model

Emerging out of the Marshall Plan's reconstruction of the European economy, the European Common Market (and soon the Common Agricultural Policy (CAP)) took a strikingly different path to that of the US agri-corporate model.

Concentrating on subsidizing the productivity of local agricultural producers, European food products were initially intended for the domestic (European-wide) market. This market was protected by significant tariffs, with subsequent elaboration of governance mechanisms such as Appellations, Geographical Indicators and notions like 'terroir' also privileging local production in the local market. When significant food surpluses emerged in the 1960s and 70s, the EU did not pursue export markets as the only solution to disposing of surplus production (although it did not entirely ignore this option either). Rather, a complex series of reforms and negotiations that both pre-dated and occurred in parallel to the GATT Uruguay Round (1985-1995), culminated in the McSharry Reforms of 1992. These represented an attempt to mitigate the full challenge of the trade liberalization that was being mobilized against European agriculture in the GATT negotiations. The tactic of negotiators to agree to a process of removal of tariff protection around the European market by some distant date also bought time to allow the elaboration of a particular farming model that could enable Europe to continue as a leading member of the WTO (World Trade Organization) while still protecting the livelihoods of its domestic producers.

The particular European model for supporting domestic agriculture that emerged from this period is often termed 'multifunctionality'. Deployed as an explicit counter-argument to the trade liberalization rhetoric at the heart of the GATT negotiations, multifunctionality is a position that recognizes multiple functions to agriculture and farming. Rather than reproducing the market logic of neoliberalism (that is, simply producing food at the cheapest possible price for the welfare of the global market), multifunctionality promotes the idea that farming not only produces particular ecological and cultural effects, but also underpins a particular culturally and politically valued configuration of the rural landscape. Using this as a guiding framework, price support subsidies (which are termed 'blue box' subsidies) could be transitioned to environmental subsidies (which are termed 'green box' subsidies). Seen through this policy approach, farmers can be subsidized to produce ecological and cultural benefits even while being technically 'liberalized' from price supports and tariff protection to produce agricultural products for the world market.

This particular model of agriculture is so explicitly targeted at securing livelihoods for domestic farmers that it has little relevance to wider discussions of solving world hunger (in contrast to the US agricultural use of world hunger as a rhetorical device to argue for increased intensification and expansion of the industrial farming model). Rather, the key contestation of the tension between securing European farm livelihoods and finding solutions to world hunger comes in issues around market access to European consumers by Developing World producers. In this context, some NGOs (non-governmental organizations) such as Oxfam have explicitly argued for further liberalization of European food markets to open up opportunities for expanding food production in the Developing World, leveraging off the returns available in a wealthy market like Europe.[2] Others respond that if Europe completely opened its markets, it is unlikely that hungry, small-scale producers in the Developing World would be the actual beneficiaries of new market opportunities, with the vastly better resourced and dominant US agri-corporates more than likely to fill the gap. None of these discussions change the basic orientation of European farm policy from one primarily directed at supporting the livelihoods of their own farmers. Put simply, Europe's policy ambition begins and ends with: 'let us preserve the producers of cake'!

### The New Zealand model

Both the US and European models received a significant and prolonged challenge from the resurgence of nineteenth century market liberalism in the 1980s. Just as the repeal of the Corn Laws had changed the parameters of world trade in agriculture mid-nineteenth century, the agreement of the US and Europe (along with other key countries such as Japan) to allow agriculture to enter discussion in the Uruguay Round of the GATT in 1985 signalled a possible step-change in global trade in agricultural products. While both the US and Europe found mechanisms to blunt the effect of potential trade liberalization, other countries – notably the 'Cairns Group' of free traders – embraced the new liberalized trade model as the way to secure sustainable livelihoods for their farmers. Foremost among these countries was New Zealand – which provides some kind of paradigm of the 'neoliberal' model of securing agricultural livelihoods.

New Zealand was not always a free trade advocate. Until the entry of the UK into the European Common Market in 1973, New Zealand – as a faithful ex-colony – had enjoyed unfettered access to the lucrative UK market. Under this privileged arrangement, New Zealand agriculture had solidified its niche in world agricultural trade via the mechanism of 'single-desk' producer boards and extensive state investment into agricultural R&D. The loss of privileged market access to the UK in 1973 created a major crisis for which the neoliberal model, apparently, provided the solution.

The rapid adoption of neoliberal policies in agriculture provoked an even deeper crisis (for a review see Campbell and Lawrence, 2003), although many commentators (including this author) have been surprised at the degree to which a medium-term recovery took shape around a particular model of high-value, quality-audited produce directed towards high-end multiple retailers in markets like Europe and the US (see Campbell, 2009).

At the heart of the New Zealand model lies a dynamic identified by Peck and Tickell (2002) as 'roll back' and 'roll out' neoliberalism. According to this schema, the first phase of neoliberal reform involves a rolling back of state regulatory functions and structures to 'liberate the market'. When this turns out to be less than successful (or blatantly dysfunctional), a 'roll out' phase takes place in which new forms of governance and organizational experiments begin to fill the gap left by the removal of state regulation. In the case of New Zealand, the roll out phase involved the proliferation of forms of private sector governance within agri-food systems (Le Heron, 2003), which reintegrated New Zealand's food export industries into wider formations of new economic and governance arrangements (Larner and Le Heron, 2004). This new style of integration into world food markets has been peculiarly powerful for those food export industries that have moved into the high-value, quality-audited niche in the world food market, resulting in entirely new agri-food assemblages as well as transformed praxis of food production at the level of orchards and farms (see Rosin, 2008; Rosin *et al.*, 2008).

At first glance, the New Zealand model for adopting high-value, quality-audited food products for elite markets seems to strongly vindicate the neoliberal model for securing livelihoods for agricultural producers. If we return, however, to the compelling need to feed the world, two provisos become apparent. First, the capacity of elite UK supermarkets to provide high monetary returns to elite suppliers in premier agricultural export nations like New Zealand surely represents a global niche rather than a model for the entire market. In fact, it is arguable that this elite niche only exists because of some of the repellent qualities of the industrial food system (see this book, chapters by McMichael, Stock and Carolan, Butler, Dixon and Pritchard). It is only the by-products of this elite model that find their way to the bottom end of the world market – particularly, in New Zealand's case, the trade in cheap, and extremely fatty 'mutton flaps' that are sold through various low-profile arrangements into the South Pacific (Gewertz and Errington, 2010). Clearly this does not represent a useful way to address world hunger. Second, there is some suspicion that these new private sector governance arrangements are actually acting as a means of excluding Developing World producers that is, if anything, even more exclusive than the prior, more state-organized style of agricultural and trade regulation (Campbell, 2005). At least if the UK government were engaging in exclusionary practices that discriminated against poor producers in the Developing World there would be a mechanism of political recourse via the WTO disputes process. If Sainsbury's does the same, however, then there is no mechanism whatsoever available to those who find themselves locked out of such market opportunities.

Like the US model of corporate, commodity-based export agriculture, the New Zealand model is often politically legitimated on the grounds that expansion of world markets for food (even if it is elite foods such as spring lamb cuts or organic kiwifruit) is the key means for solving world hunger. In the case of New Zealand, rather than 'let them eat cake', the phrase should be recast as: 'let them eat our organic cheesecake with kiwifruit jus'!

## Solving world hunger and the multiple projects sustaining the livelihood of Developed World farmers

One of the greatest – and also most ambiguous – achievements of the Industrial Revolution was to effectively end the problem of food scarcity in the industrial countries (if not the problem of inequality of access). Twice in the history of industrialization there have been periods where this has been challenged. In the 1840s (the 'Hungry Forties') the rapidity of population growth in Britain outstripped the ability of domestic farmers to supply the local market. The solution to this period of significant destabilization was the repeal of the Corn Laws and a transport revolution that created a global-scale colonial food system to feed the burgeoning population of the imperial/industrial core countries. This system retained some degree of stability for the better part of a century, representing what Friedmann and McMichael (1989) called the First Food Regime. Again, in the middle of the twentieth century, the Great Depression, Second World War shortages and rationing, and the menace of the Cold War disrupted the ability of core industrial countries to feed their populations. This prompted a major restructuring of world food relations, with the Second Food Regime turning the focus of agriculture and food policy away from far-flung colonies and squarely back in the direction of domestic food producers in the industrial core. This second major shift was justified in terms of the new need for 'food security' at home, just as the bread riots in the English Midlands had been used to justify the repeal of the Corn Laws (and the rape of Ireland) in the Hungry Forties a century earlier.

In both these cases, it was not the presence of global hunger that prompted any shift or transformation of political or policy responses. Rather, it was the possibility of disruptive hunger and starvation in the industrial countries themselves that prompted a dynamic restructuring of global food relations – in the first case to create a global-scale system of colonial food supply and in the latter to dramatically reverse it. An important question is what kind of precedent this might set for another major revolution in the nature of the global food system? After all, we clearly need such a revolution if we are actually going to solve world hunger in a sustainable way.

The immediate response is that the lessons from history are not as encouraging as they might first appear. The basic insight that the world food system has already gone through two tumultuous periods of upheaval and reconstruction provides some grounds for hope that it may happen again – this time with more enduring positive outcomes. However, closer inspection of the historical precedents is not quite so inspiring.

First, these upheavals took place with scant regard for actually solving the problem of hunger in the non-industrial world. Mike Davis's account of the creation of world food markets along the lines of empire (in his case, in India) clearly linked this 'solution' to the food supply crisis in the industrial core with the simultaneous deaths of tens of millions of peasants in the colonies (Davis, 2001). Similarly, the post-Second World War return towards privileging domestic producers in the cause of food security in the Developed World, and the various political projects of the Second Food Regime post-Second World War, seemed to have no positive impact on reducing the number of starving people in the world. The ambiguous outcomes of the green revolution and the intensification of agriculture in the Developed Countries were successful only if viewed as a mechanism to support farm incomes in the Developed World, as well as stimulating the economic development of global agribusiness. While threats to the food security and stability of food supply in the industrial countries certainly acted as the spur to major changes in food systems, solving world hunger was not the explicit or intended target of the resulting responses.

Second, while history demonstrates some important cases of dramatic reconfiguration and experimentation, the kinds of resulting models for agriculture do not pose a useful solution for solving world hunger – even if the architects of these solutions sometimes hoped, or rhetorically suggested, that solving world hunger was part of the deal. The US model – particularly the export-oriented, aid-leveraged model of the Second Food Regime – has been credited with widespread destruction of local food systems in the Developing World; the migration of hundreds of millions of peasants to the slums of Developing World mega-cities; the creation of elite food 'export platforms' that provide new economic opportunities only for those with capital to invest (none of whom are actually hungry); and for establishing a style of industrial agriculture that is widely blamed for creating the conditions for ongoing ecological unsustainability of agriculture.

In contrast, the European multifunctional model attempts, at least, to address economic, ecological and cultural dynamics of agriculture simultaneously. In this sense, the multifunctional model provides the closest point of purchase between idealized Developed and Developing World models of agriculture. However, it is also extremely inward-looking and is clearly aimed at reproducing a particular European style of agriculture. The ability of this model to operate successfully in a more international context is surely questionable in the absence of a large, wealthy domestic market and a prosperous polity that can afford a high level of direct and indirect subsidization.

The neoliberal model exemplified by New Zealand is often cited as a potential model for wider development of economically sustainable global agriculture within the context of a global free market. The absence of subsidies and a highly export-oriented agricultural economy makes New Zealand superficially more like a Developing World agricultural producer than the US or Europe. That, however, is where the similarity ends. New Zealand has prospered under neoliberal conditions because of extremely propitious, temperate production conditions; a highly educated agricultural workforce; high levels of literacy and computer-use; politically empowered grower organizations; strong direct dialogue with purchasing agents from elite multiple retailers; marketing expertise; a relatively highly resourced agricultural R&D sector and a very positive global brand image. Yes, New Zealand has no subsidies, but it is not a Developing World country. Accordingly, it is bordering on the delusional to assume that the neoliberal model demonstrated by New Zealand has relevance to Developing World contexts where actual participation in the formal market is by no means guaranteed.

## Conclusion: if cake is not the answer, then what is?

This longer term examination of food since the Industrial Revolution tells us some important things about transformative possibilities in world food relations. First, we are not 'locked in' to the existing system: there are conditions under which major transformation can take place. Twice in the last 170 years, the world food system's fundamental structure has been radically altered. In both cases, food security concerns in the wealthy nations drove that change. The coincidence of both these transformations taking place during the fifth decade of each century should not be given too much attention, as it is not only possible but highly likely that we may not be waiting until the 2040s before such transformative conditions arise again. Many of the chapters in this volume identify conditions like instability in world futures markets, peak oil and climate change, as dynamics that may, quite soon, destabilize the food supply of wealthy countries in the same way as briefly occurred in 2008 and more compellingly in the 1840s and 1940s.

When that does happen, it is essential to recognize a second key lesson from our recent history: namely, that the current three prevailing models for organizing agriculture in the Developed World provide few clues for transforming food security and the sustainability of food systems. After all, this was not their purpose. They were designed to maintain the incomes of Developed World farmers and each, in its own way, has succeeded in that goal. They have enabled the Developed World to continue to eat cake.

So, if our prevailing Developed World models of agriculture are not providing a useful template for sustainable global food security, do we just abandon hope? The answer is clearly no! Just because these models turned out to be finely tuned mechanisms for creating incomes for Developed World farmers does not mean that the compelling challenge of food security has not been addressed from elsewhere. There is more to the world of food than that encompassed by the aspirations of the US, European and New Zealand models of agriculture. Part of the problem lies in the very notion of a broadly applicable 'model' of agriculture. In a recent paper, McMichael (2010) argued for a recognition of the dynamic process of 're-peasantization' of global agriculture as a key contributor to the future sustainability of global food relations. This argument brings together the collective efforts of all forms of agriculture which seek to create conditions under which the social, economic and ecological reproduction of agricultural production forms can survive. Re-peasantization is not about creating a new global 'model' for agriculture, it is about recognizing the basic ingredients of long term reproduction of agriculture and recognizing those ingredients when they occur in a range of settings – from both the Developing and Developed World. There is an important reason why these kinds of solutions will come in a multiplicity of forms. If the challenge – seen in McMichael's (2010) terms – is to foster agriculture that is socially, ecologically and economically sustainable, then the challenge of finding locally-appropriate ecological relations for agriculture will not be well served by 'one size fits all' global-scale models for agriculture. Rather, we need to foster spaces of experimentation and multiple pathways for development of food security.

The grand narratives of agricultural intensification, productivity, liberalization and multifunctionality all perform excellent service as the central motifs of large policy projects at a global scale. They are less useful, however, in capturing the multiple contingencies and specificities of locally sustainable agriculture in a myriad of regional and local food systems. My key contention is that these grand narratives must shrink and recede in order to open up these other spaces where experiments in food security can take place.

This is both a pragmatic and a political claim. First, pragmatically, it appears that we have reached some kind of high water mark of the reach of a globally integrated market for food. The recent rises in food prices (two now since this book was conceived) suggest an end to one of the neoliberal project's key tenets: that an integrated world food market would expand on the back of ever-decreasing prices for food commodities. The end of this long decreasing trend in food prices now presents a profound challenge to this fundamental tenet of market-based solutions to global food security. The world food market will certainly not expand in an age of increasingly expensive global food commodities. A second key trend is that we are also entering the declining years of a global energy regime that provided cheap oil-based transportation to facilitate the operation of a world food market. The approaching end of the age of oil will contribute to the upcoming contraction of the scale and reach of the global market for food.

Even given just these two long-term trends, there are clear pragmatic reasons why new space will open up at (or outside) the margins of the globally-integrated market for food. Inside this space, McMichael (2010) has already identified a range of political projects and experiments that have the potential to transform food security in multiple, specific contexts. Taking into account that these experiments are taking place in specific ecological conditions, often at a small scale, and with the kind of dynamic economic experimentation that characterizes much business at the margins, they hold considerable promise for maintaining and potentially improving food security in these contexts. History has witnessed one major flip in the orientation of global food from colonial supply to domestic food security. Perhaps we already have sufficient evidence to suggest that another flip is now required, from the existing models in the Developed World to the enduring systems of the Developing World.

## Notes

1  A historical dynamic that would interest historians from Thompson (1963) to Braudel (1973) and Hobsbawm (1984).
2  As it stands, the EU has already committed to a timetable for tariff reduction under its GATT Uruguay Round obligations.

# References

Braudel, F. (1973) *Capitalism and Material Life: 1400–1800,* Harper & Row, New York

Campbell, H. and Lawrence, G. (2003) 'Assessing the Neo-liberal Experiment in Antipodean Agriculture', in R. Almas and G. Lawrence (eds) *Globalization, Localization and Sustainable Livelihoods*, Ashgate, Aldershot, pp89–102

Campbell, H. (2005) 'The Rise and Rise of EurepGAP: The European (Re)Invention of Colonial Food Relations?', *International Journal of Sociology of Agriculture and Food* vol 13, no 2, pp6-19

Campbell, H. (2009) 'Breaking new ground in food regimes theory; Corporate environmentalism, ecological feedbacks and the 'food from somewhere' regime', *Agriculture and Human Values,* vol 26, no 4, pp309-319.

Davis, M. (2001) *Late Victorian holocausts: El Nino Famines and the Making of the Third World*, Verso, London

Friedmann, H. (1990) 'The Origins of Third World Food Dependence' in, H. Bernstein, B. Crow, M. Macintosh, and C. Martin (eds) *The Food Question: Profits Versus People*, Monthly Review Press, London

Friedmann, H. (1993) 'The Political Economy of Food', *New Left Review,* vol 197, pp29-57.

Friedmann, H., and McMichael, P. (1989) 'Agriculture and the state system', *Sociologia Ruralis,* vol 29, no 2, pp93–117

George, S. (1976) *How the Other Half Dies* Penguin, Harmondsworth

Gewertz, D. and Errington, F. (2010) *Cheap Meat: Flap Food Nations in the Pacific Islands,* University of Calilfornia Press, Berkeley, CA

Hobsbawm, E. (1984) *Worlds of Labour*, Weidenfeld and Nicolson, London

Larner, W. and Le Heron, R. (2004) 'Global benchmarking: Participating "at a distance" in the globalizing economy', in W. Larner and W. Walters (eds), *Global Governmentality: Governing International Spaces*, Routledge, London, pp212-232

Le Heron, R. (2003) 'Cr(eat)ing food futures: reflections on food governance issues in New Zealand's agri-food sector', *Journal of Rural Studies,* vol 19, no 1, pp111-125

McMichael, P. D. (2010) 'Food Regimes Transitions', paper for the RC-40 plenary at the *World Congress of Sociology*, Goteburg, July 11–17, 2010

McNeill, J. (2000) *Something New Under the Sun: An Environmental History of the 20th Century*, Allen Lane, The Penguin Press, London

Mintz, S. (1985) *Sweetness and Power*, Elisabeth Sifton Books, New York

Peck, J. and Tickell, A. (2002) 'Neoliberalizing space', *Antipode,* vol 34, pp380-404

Rosin, C. (2008) 'The conventions of agri-environmental practice in New Zealand: farmers, retail driven audit schemes and a new Spirit of Farming', *GeoJournal,* vol 73, pp45-54

Rosin, C., Campbell, H., and Hunt, L. (2008) 'Audit Me This! Orchard-Level Effects of the EurepGAP Audit System on New Zealand Kiwifruit Producers', in C. Stringer and R. Le Heron (eds*) Agri-food Commodity Chains and Globalising Networks*, Ashgate, Aldershot

Salaman, R. (1949) *The History and Social Influence of the Potato,* Cambridge University Press, Cambridge

Thompson, E. P. (1963) *The Making of the English Working Class,* Gollancz, London

Trollope, A. (1863) *North America Vol 1,* J. B. Lippincott & Co, Philadelphia

Wolf, E. (1982) *Europe and the People without History*, University of California Press, Berkeley, CA

# 4

# TRADING INTO HUNGER?
# TRADING OUT OF HUNGER?
# INTERNATIONAL FOOD TRADE
# AND THE DEBATE ON
# FOOD SECURITY

*Bill Pritchard*

The need to generate a sophisticated global policy framework to ensure the complementarity of international trade and food security should represent one of the most pressing international programmes of the current age. Yet the brutal reality is that there is very little in the way of an accepted set of principles or policy spaces to advance this agenda. Any semblance of a debate in this topic area is terminally fractured by schisms of ideology, disciplinarity, communities of interest (some with vested commercial concerns) and institutional fragmentation. Depending on who a policy-maker speaks to, the ability to trade food across countries and continents can be portrayed either as the answer or an obstacle to improved food security.

This chapter seeks to explain this divergence of opinion. It contends that the heart of the problem is an institutional disconnection between the multilateral rule-setting regime for food trade and commerce (primarily through the World Trade Organization (WTO)), and that for monitoring and promoting food security (primarily through the Food and Agriculture Organization (FAO)), and the Special Rapporteur on the Right to Food, which operates through the UN High Commissioner for Human Rights). What this means is that, on the one hand, the WTO-based global trade agenda is advanced with insufficient regard to food security, whilst on the other, efforts by the FAO and the Special Rapporteur to advance food security are circumscribed by the overarching legal requirements of countries to abide by WTO rules. The resultant policy mishmash ill serves the needs of the world's hungry.

To appreciate the background and implications of this institutional disconnection, this chapter firstly discusses the role of the WTO, with particular reference to its Agreement on Agriculture (AoA), the primary document detailing the rules of international food trade. The chapter then outlines the concept of food security, with regards to the ways it is understood and advanced by the FAO and the Special Rapporteur. The third major section of the chapter details the disconnection between these two combative approaches and the global politics of food. Finally, the chapter asks whether – and how – the current set of arrangements can, and should, be altered. The point is that the world deserves a more inclusive, high-level conversation (across disciplines, civil society, governments and the private sector) on this topic, connected to a new institutional architecture of international rule-setting around food. The crude certainties that typify so much of (what passes for) international debate on trade and food security in the WTO era (a set of black-and-white perspectives arguing that trade is good/bad for food security) warrant replacement by a more contextualized approach that calibrates trade rules and policy to understandings of how exposure to international trade generates resilience and/or vulnerability among members of food-insecure populations.

## International food trade and the World Trade Organization

The WTO is an unusual international organization. It is constituted by a series of Agreements established in the mid-1990s that specify limits to which members can intervene in global trade. Yet at the same time, its mandate is to renounce those same Agreements by working towards newer ones which further liberalize world trade. This contradiction in the WTO (as upholding existing Agreements yet working to replace them) fundamentally shapes its relationship to the debate on international food security.

The goal of having a worldwide set of rules to govern trade can be traced to 1941, when Winston Churchill and Franklin D. Roosevelt established the Atlantic Charter: a statement of economic and political principles on which their two countries would base their ongoing relations. Principle four of the Charter specified that the two countries would 'endeavour, with due respect for their existing obligations, to further the enjoyment by all states, great or small, victor or vanquished, of access, on equal terms, to the trade and to the raw materials of the world which are needed for their economic prosperity'. These innocuous sounding commitments provided the foundation for the following 65 years of government-to-government regulation of international economic activity. Initially, they provided the basis for the General Agreement on Tariffs and Trade (GATT), the agreement penned over the course of meetings in 1945–47 for the ultimate purpose of providing a constitution for a proposed International Trade Organization (ITO). Intended to be part of the United Nations family of international organizations, the putative ITO was abandoned in 1948 following objections from the US Congress. Nevertheless, the GATT, drawn up by a small cadre of 15 countries, and with a limited scope and purpose, lived on until 1994.

The eventual demise of the GATT regime was instigated by the decision to commence multilateral trade talks in Punta del Este, in Uruguay, in 1986. The

Uruguay Round, as it became known, involved a series of wide-ranging negotiations which focused on the alleged shortcomings of the GATT system. For the purposes of this chapter, there were two key shortcomings identified. The first of these related to the issue of discipline and enforceability. The GATT was a loose, treaty-based regime without a robust and permanent system for dispute settlement. As such, its viability depended on members' commitments to abide by its provisions. As the GATT's membership steadily grew, and international trade increased in volume and complexity, the problem of membership commitment and enforceability became increasingly vexed. Thus, by the mid-1980s, there was consensus amongst trade policy elites for a replacement multilateral regime, based around a permanent institution with powers to enforce decisions on members – in other words, a return to the original 1947 conception of an ITO. The second key issue related to agriculture. This sector was excluded from the GATT's core provisions, largely on account of US domestic policy concerns in the late 1940s. During the GATT era, therefore, international agricultural trade negotiations tended to take place via purpose-specific 'side-agreements' (the International Dairy Agreement, the Bovine Meat Agreement, etc.), generally involving limited numbers of the GATT membership. Such arrangements clearly fostered institutional complexity in the international trading system: whether a country was a signatory, or not, to any Agreement determined its behaviour on world markets. Non-signatories could episodically dump product into others' markets with no effective sanction, thus injecting considerable uncertainty and unpredictability into international trade. Moreover, the disjointed nature of this system meant that agendas to reform agricultural trade had little traction, as there was no single forum for discussion of sector-wide issues. During and following the course of the Uruguay Round, the argument for the inclusion of agriculture into the core body of a single multilateral trade system was pursued with vigour by a coalition of agricultural export nations called the Cairns Group, chaired by Australia.

The seven years of Uruguay Round negotiations culminated in a decision to replace the GATT with the WTO. This created a single institutional entity, not only with enforceable dispute settlement powers but also a mandate that included agriculture. The latter was institutionalized through the Agreement on Agriculture (AoA), a document which established a set of binding obligations on members to limit – and progressively reduce – the support they provided to their agricultural sectors. These obligations covered three broad areas. Firstly, support designed to restrict *market access,* in the form of tariffs and quotas which make imports less competitive. Secondly, support in the form of *domestic payments* by governments to farmers above and beyond what they receive from the market sale of their products. This has proved a highly contentious area of negotiation and dispute due to the diversity of rationales and mechanisms through which such payments are made. In the WTO, key distinctions are made between payments which distort markets because they allow farmers to sell their output at below-market prices ('amber box'); those that distort markets but have quantitative limits on payments, thus reducing their market-distorting impacts ('blue box'); and payments which are defined as being non-market distorting (or 'green box'), such as compensatory payments for natural disasters, or for farmers to undertake legitimate remedial environmental actions. Finally, the AoA specifies a set of obligations on members with regards to a third type of agricultural support, which is *export subsidies*. These result where governments offload surplus production by allocating subsidies which allow this surplus to be sold on world markets at prices below the cost of production. It is generally regarded that export subsidies are the most damaging of all kinds of agricultural supports, because they facilitate the large-scale dumping of product onto world markets, with particularly serious effects in terms of their capacities to disrupt smaller, vulnerable producer communities.

The incorporation of agriculture into the core of the multilateral trading system has had significant impact on the global food system, but in ways not necessarily anticipated at the time the WTO was created. In the early 1990s, many analysts assumed that the inclusion of agriculture in the multilateral system would lead to significant and immediate reductions in protection within developed countries (Le Heron, 1993; Friedmann, 1994). Such assumptions, however, proved ill-founded. The best way to demonstrate this is to examine official statistics of the 'Total Support Estimate' to agriculture in the OECD (Organization for Economic Cooperation and Development), expressed as a percentage of total farmgate production (OECD 2010). In 1995, this was 46.93%. Ten years later, in 2005, it had barely changed – it was 44.26%. In the following few years it declined substantially (reaching a low of 33.91% by 2008); but, crucially, rather than being due to subsidy reductions (i.e. reductions in the numerator) this was due to the effects of escalating agricultural prices in boosting the denominator. In 2009, after the food price bubble burst, the total estimated level of support to agriculture within the OECD climbed again, to 38.67% of farmgate sales.

The slow rate of progress in reducing agricultural subsidies in the developed North occurred because of three intervening processes. First, the difficulty of reaching consensus in the Uruguay Round meant that the AoA ended up specifying only a relatively modest timetable for reductions in agricultural tariffs and subsidies. Liberalizers (such as the Cairns Group) begrudgingly accepted this outcome because they regarded the inclusion of agriculture within the multilateral system as a 'win' in its own right. Second, it was widely presumed in the mid-1990s that the conclusion of the Uruguay Round would be followed quickly with the commencement of a new Round. However, after false starts the eventual Doha Development Round commenced in October 2001 – fully seven years after the Uruguay Round's completion. By this time, the multilateral system was bruised from institutional failures and an increasingly sceptical public (as witnessed in the abortive 'Millennium Round' meetings in Seattle, 1999). The Doha talks were consequently undertaken within a very restrained set of aspirations. Third, developed countries adopted an array of strategies to subvert the (already modest) requirements of the AoA. With regards to tariffs, developed countries calibrated their reduction targets to artificially high baselines, meaning that they could meet their obligations with relative ease (a problem which became coined as 'water-in-the-tariff'). With regards to domestic payments, developed countries engaged in an elaborate process of 'box-shifting', whereby they restructured their farm subsidies from 'amber' and 'blue' to 'green' categories. Thus, in the EU, a process known as 'decoupling' saw the abolition of many direct production payments and the creation of new, non-market distorting farm payments designed to promote environmental services, heritage and rural amenity. These 'Pillar 2' payments have come to represent a key component and discursive justification for why the EU engages in agricultural support activities. As suggested by Clive Potter, only in the past decade has the reshuffling of EU subsidy payments to ensure they are 'WTO-consistent' cohered as a discrete political project. As Potter (2006, p192) further explains:

> negotiators have invented a complicated new vocabulary which establishes procedures for tarification and creates a system for estimating and classifying domestic subsidies according to the degree of 'trade distortion' they imply. A paradoxical consequence of this is that as a champion of deregulation, the WTO has found it needs to invent complicated new vocabularies and legalistic procedures in order to achieve its objective.

These outcomes have been particularly invidious for the efficiency and equity of the global food system. By allowing the world's rich nations to persist with high levels of agricultural support, the AoA exposed developing countries to subsidized agricultural imports from Europe and the US, thus undercutting the competitiveness of local production (a problem of market distortion), and also restricted their ability to sell agri-exports into affluent developed countries (a problem of market access). The net result of these processes was to generate the perverse situation where food-secure Northern countries further expanded their production of basic food staples (wheat, rice, maize etc.) and relatively food-insecure Southern countries switched to luxury, high-value crops (tropical fruits, floriculture etc.) (Fold and Pritchard, 2005). Hence, the net effect of the WTO was to institutionalize a set of imbalances in the global agri-food trade.

The WTO's incongruities sit uneasily both within and outside the organization. Its defenders emphasize its role in enshrining international rules in agricultural trade, and working towards liberalization. Yet both these arguments contain contradictions. The WTO upholds the AoA – yet, as already discussed, this is a flawed agreement, and one that has had very limited success with regards to progressing agricultural trade liberalization. Developed country members have met their AoA obligations by hook or by crook, and, in a number of instances (including subsidies to European sugar producers and US cotton growers), only after the WTO has been forced to rule against these countries in lengthy, high-profile dispute settlement cases. Even then, after rulings have gone against Northern subsidies, subsequent 'box-shifting' initiatives have muddied the waters and it is not clear at all that developing countries have benefited. Indeed, after fifteen years of existence, there is only one instance of a substantial, multilaterally-endorsed decision to liberalize beyond the frames of the AoA. This occurred at the Hong Kong WTO Ministerial meeting of 2005, when members agreed to wholly eliminate export subsidy payments by 2013. As earlier noted, the highly distortive nature of these payments means this form of subsidy can wreak havoc on markets, and so their elimination was universally welcomed. However, as Oxfam noted at the time, because these subsidies actually form only a small proportion of total Northern agricultural subsidies, the quantitative impact of their elimination will be slight (Oxfam, 2005).

To summarize, the WTO is a member-driven organization which was established on the basis of a series of compromises at the end of the Uruguay Round. By institutionalizing these imperfections into the multilateral rules of agricultural trade, it has acted to aggravate social and economic contradictions in the global food system. The modest levels of subsidy reductions in the first fifteen years of the organization speak to the overall resistance in the developed North to the pursuit of change. These countries have managed to retain support for their farming sectors through various strategic 'box-shifting' mechanisms. Moreover, attempts to break this impasse via the creation of the Doha Round of multilateral negotiations proved inconclusive. The collapse of these discussions in 2008 marked the end of a multilateral momentum within the WTO (Pritchard, 2009). When taken together, when it comes to the WTO, all these issues point to a gaping hole between purpose and achievement. As time has passed, its incapability to resolve the tensions of global agriculture has become increasingly apparent. This context being set, we now turn our attention to the second of the key foci of this chapter – the concept of food security.

## Food security

The concept of 'food security' is much misunderstood. It is often erroneously conflated with notions of food self-sufficiency: whether a community (a nation, city or region) has enough own-production and buffer stocks to feed its population over a given period. However, in contemporary international debate, food security has evolved to develop a meaning which is more expansive than just self-sufficiency. The intention of this section, therefore, is to set out this meaning and, hence, provide a basis to link the concept to trade policy issues in the following section.

Until the 1990s, the prevailing perspective in international policy and research was to explain hunger in terms of the 'food availability decline' (FAD) model. This was a production-centric explanation which saw hunger as occurring when events (droughts, war etc.) disrupted food supplies and diminished food stocks. Hence, in 1974, the World Food Conference defined food security as: 'availability at all times of adequate world food supplies of basic foodstuffs to sustain a steady expansion of food consumption and to offset fluctuations in production and prices' (cited in FAO, 2006, p1).

Clearly, there is no doubt that such events can and do have profound implications for levels of hunger; cataclysms caused by weather or human action
obviously impact on people's abilities to feed themselves. However, during the
1980s there was a dramatic shift in researchers' understandings of the causes
of hunger. The 'entitlements' approach, advanced by Amartya Sen (1981),
explained hunger not simply through the prism of food production, but on the
political structures that connected people to food. For Sen, a cataclysm such
as drought or war is not necessarily a cause of hunger, but a production shock
that *triggers* hunger because of the way it may undermine people's entitlement
arrangements.

This approach reflected a radically different lens into this issue. Sen's research
into the Bengal famine of 1942-44 overturned prevailing official and scholarly
explanations. He saw a 'moderate' fall in production being translated into an
exponential number (perhaps three million) of hunger-related deaths because of
the specific nature of prevailing social and political arrangements. These included,
firstly, trade restrictions on rice which prioritized the maintenance of food stocks
in Calcutta over and above rural areas of Bengal and, secondly, war-related food
price inflation that inhibited landless rural labourers' abilities to access food.
Hence, rather than being a famine afflicting an entire region's population, it was a
phenomenon with sharp social and economic cleavages. Sen argued that dramatic
changes to the politics of food in Bengal during 1942-44 effectively disen-
franchised food access for the rural poor, leading to his characterization of the
phenomenon as a 'boom famine' (1981, p75). It was not food deficit that caused
starvation, but changes in the political arrangements that connected vulnerable
individuals to the food system.

From this analysis, Sen proposed the concept of *entitlements* as a generalized
way of understanding hunger. This concept sought to explain hunger and famine
by way of asking questions about the social, cultural and economic frameworks
that bestowed rights to ownership within populations. As Sen contended, even
the seemingly straightforward notion of possessing a loaf of bread (and thus fore-
stalling one's hunger) assumes a chain of entitlement relations:

> I own this loaf of bread. Why is this ownership accepted? Because I got it
> by exchange through paying some money I owned. Why is my ownership of
> that money accepted? Because I got it by selling a bamboo umbrella owned
> by me. Why is my ownership of the bamboo umbrella accepted? Because I
> made it with my own labour using some bamboo from my land. Why is my
> ownership of the land accepted? Because I inherited it from my father. Why
> is his ownership of that land accepted? And so on. Each link in this chain
> of entitlement relations "legitimizes" one set of ownership by reference to
> another, or to some basic entitlement in the form of enjoying the fruits of
> one's own labour.
>
> Sen (1981, p1–2)

Thus, the ability of a person to avoid starvation depended on her/his entitlements, which, in turn, were constructed from her/his *ownership bundle* (the combination of labour powers, resources and assets s/he can use to acquire food) and the *exchange entitlement regime* s/he faced (the rights to resources that can be accessed to transfer an ownership bundle into food).

The entitlement approach was significantly influential at the 1996 World Food Summit (WFS), where the FAO redefined food security in terms notably distant from its previous, production-centric (emphasizing food stocks) logic: 'Food security exists when all people, at all times, have physical, social and economic access to sufficient, safe and nutritious food that meets their dietary needs and food preferences for an active and healthy life' (FAO, 1996).

The overriding message in this definition is that the attainment of food security hinges on the social, economic, cultural and political circumstances that either enable or restrict the provisioning of food to needy populations (Levendal *et al.*, 2004; Chung *et al.*, 1997). This orientation gave rise to a framework which held that food security was the outcome from three sets of processes:

- Food availability: the supply-side factors which shape the availability of sufficient quantities of food of appropriate quality.
- Food access: the political, social, cultural and economic processes that connect supply-side processes to individuals, and
- Food utilization: the elements of clean water, sanitation and health care that ensure that food that is made available and is accessible (i.e. the two categories above to generate nutritional well-being for consumers).

This framework has crucially informed the work of key international organizations since the 1990s, with important ramifications for their advice to developing countries. The FAO, notably, has widened its policy interests in this field during the past decade. Whereas traditionally this organization was focused on generating international policy capacities and coherence around the issue of global monitoring of food production and stocks, it now supplements these interests with heightened interests in issues relating to food access and utilization (the latter through its involvement in *Codex Alimentarius* – an international food standards-setting body it runs jointly with the World Health Organization). This broader sphere of interest is seen readily in its annual *State of Food Insecurity* reports, which explicitly adopt the entitlements framework to explain the changing geographies of undernourishment.

Complementing these foci are initiatives relating to the Right to Food. This right was enshrined in the UN Universal Declaration of Human Rights in 1948. Article 25(1) of the Declaration states:

Everyone has the right to a standard of living adequate for the health and well-being of himself and of his family, including food, clothing, housing and medical care and necessary social services, and the right to security in the event of unemployment, sickness, disability, widowhood, old age or other lack of livelihood in circumstances beyond his control.

The translation of the sentiments of this statement into a more concrete platform for international obligations has occurred over decades, and has involved various different organizational arms within the UN. Since the turn of the millennium, two developments have been key to progress. Firstly, in the year 2000, the UN Commission on Human Rights agreed to establish an office of the 'Special Rapporteur on the Right to Food'. In general terms, the mandate of this office was to promote international cooperation and information on members' progress with respect to the rights to food. Then, in 2004, the 127th session of the FAO Council endorsed a set of 'voluntary guidelines' which member states could use in order to incorporate these rights into national legislation (FAO, 2004). The influence of these initiatives has been considerable in the ways that food security has been framed within international forums. Effectively, they have solidified the entitlements-based perspective by connecting it to a set of international legal rights.

India provides an excellent example of the influence of these perspectives on food security policy. Since independence, India has operated a food safety net policy known as the public distribution system (PDS). In most states across India, the PDS operates in tandem with poverty line measurements, so that households deemed to be below poverty line (BPL) gain an entitlement to highly-subsidized basic foodstuffs. Historically, however, considerable problems of maladministration and leakage have dogged this system; problems which in the late 1990s led to public interest litigation against the government of India for failing to ensure the rights to food for its citizenry. In a 2001 decision, the Supreme Court of India ordered the establishment of an Office of Commissioners charged with monitoring India's compliance with the right to food. Then, at the 2009 national election, the ruling Congress Party announced an intention to legislate for the right to food, thereby supplementing the PDS with a legally-binding right for citizens to be provisioned with a determined minimum level of food staples. At the time of writing, the final form of this legislation is yet to be determined. However, whatever shape it eventually takes, it will mark a significant shift in Indian food policy.

To summarize this section, in post-1990s international forums, the prevailing view of food security is centred on the concepts of entitlements and rights. The entitlement approach, flowing from the work of Amartya Sen, has augmented traditional supply-side (production and stocks) perspectives on food security by foregrounding the political structures that shape food access. Greater formal recognition of individuals' rights to food within international and national law has complemented this focus.

## Disconnections between food security and the multilateral regulation of international trade

As noted in the introduction to this chapter, there is an ongoing failure of the global community to generate a framework which ensures the complementarity of international trade and food security policies. The two previous sections have set out the basic reasons for this disconnection. The review of the WTO revealed it as administering a flawed set of global trade rules for food (the Agreement on Agriculture), and also possessing an agenda to replace this Agreement with one that further liberalizes world food trade. From the WTO's perspective, these activities intersect with food security debates in terms of supply-side processes. State-imposed distortions to international food markets (via tariffs, subsidies and so on) are perceived as efficiency-impairing, and so their removal would assist food production at lower resource costs and, therefore, would potentially lower the price of food to consumers. At the same time, the section above – reviewing food security – highlighted the recent shift in focus among international organizations in this field from supply-side approaches to entitlements-based and rights-based approaches.

The appropriate metaphor here might be the 'cone of silence'. What seems to be occurring at the international level is a consideration of the relationship between food security and trade in separate discussions, premised on separate sets of assumptions, with all-too-infrequent arenas for cross-fertilization. We assess this issue by considering the failure within the WTO system to comprehensively incorporate food security concerns within its negotiating framework.

Fundamentally, the WTO's consideration of food security is ideologically subsumed within the overarching premise that trade liberalization per se improves resource efficiency, which in turn aids economic growth and improved living standards. Hence, the goal of improving global food security is best pursued by implementing freer world trade. Former WTO Director-General Supachai Panitchpakdi summed up this case as follows:

> one of the most concrete ways which the WTO can contribute to improving food security is by providing the opportunity to raise income levels through economic growth. As is recognized in the Rome Declaration and Plan of Action – trade is a key element for food security – as it stimulates economic growth. It permits the efficient transfer of food supplies from surplus to deficit regions. It allows countries to become self-reliant rather than trying to become self-sufficient, regardless of cost.
>
> WTO (2005)

On subsequent occasions, whenever WTO officials have spoken at international food security meetings, these sentiments have been consistently repeated. The WTO's strong 'take-home message' is that if trade is liberalized, then the global severity of food security problems will be alleviated. A revealing example of how the concept of food security is discursively submersed within the over-arching agenda of trade liberalization is provided in the 'glossary' section of the WTO website, where the official FAO definition of the concept (reproduced in the section above) is tweaked in such a way as to make the point that trade liberalization is not antagonistic to food security. The glossary definition notes: Food security and self-sufficiency are not the same, and a key debate is whether policies aiming for self-sufficiency help or hinder food security' (WTO, 2010).

Yet interestingly, behind the public rhetoric of WTO officials, concerns by WTO members over the relationships between food security and trade liberalization have been a steady refrain within the course of agricultural trade negotiations during the past decade. WTO members have voiced considerable and varied concerns about the implications of trade liberalization to their own food security. The most notable example of this occurred during the July 2008 Geneva Ministerial meeting, which saw Indian chief negotiator Kamal Nath steadfastly reject proposed liberalization of the Agreement on Safeguards because it would leave India open to import surges of highly-subsidized agricultural products from the EU and US. In a context where food security for the Indian population depended crucially on livelihood prospects for small farmers, Nath famously said: 'I am willing to negotiate commerce. I am not willing to negotiate livelihood security' (see Pritchard, 2009, p304).

The conflicted and unresolved nature of debate in WTO agriculture negotiations over food security reflects a fundamental weakness within the system. The problem is that the WTO system gives minimal scope for the inclusion of food security considerations within its negotiating platforms. Although the concept of food security is explicitly mentioned in the AoA's text, its meaning is limited, and certainly not consistent with the entitlements-based or rights-based interpretations discussed above. Section 3, Annex 2 of the Agreement specifies that governments are entitled to make expenditures for the 'food security purpose' of accumulating food in line with legislative requirements to maintain minimum national food stockholdings – so long as such purchases are made at market prices and undertaken transparently. The following section of the AoA then specifies that governments are additionally entitled to make purchases for domestic food aid, which involve the disbursement of food at zero or subsidized rates to eligible (i.e. nutritionally disadvantaged) members of the population.

This text in the Agreement sanctions activities such as India's PDS, which involve government purchasing/stockpiling of food grains and their subsequent distribution to needy populations at subsidized prices. Judged from a 21st century perspective, however, this is a very narrow understanding of food security. It pays no regard, for example, to the potential interactions between wider AoA compliance obligations of members and food security outcomes. A detailed, 23-country comparative review of these interactions undertaken by the FAO (2003) pointed to a hugely complex set of relationships between requirements to liberalize markets on the one hand, and food security outcomes (defined through an entitlements approach) on the other. A synthesis of findings indicated:

The most likely groups to benefit from the reduction of trade barriers in foreign markets and the expansion of exports are commercial producers. Small farmers may not be able to participate in growing export-oriented crops and may experience greater competition in accessing resources, including land, marginalizing their position even further. The winners and losers of open trade policies are likely to be different, and it is feared that it is often the poor who are hurt most. For example, the Peruvian case-study noted that fresh and processed asparagus, which emerged as the most dynamic export crop in the 1990s, is produced mainly by large farms. Greater export opportunities may also lead to the reallocation of land and other resources away from domestic food production, with possible adverse consequences for household food security.

FAO (2003)

There is minimal 'negotiating space' within current WTO agendas for such considerations to be included within the emergent scope of international trade rules. As noted above, the ideology behind WTO-led liberalization would see such issues as indirect distributional consequences of trade, which it is the responsibility of national governments to address. Employing a Ricardian comparative advantage perspective, a liberalizer would argue (using the above-cited case of Peru) that the export of fresh asparagus provides an addition to Peru's national income, which could be taxed, and the revenues from this could be used to help fund a larger food safety net, thus producing a win-win outcome for both the national economy and food-insecure households. Such a model is not to be dismissed out of hand – international statistics demonstrate overarching correlations at the national scale between economic growth and net reductions in undernourishment – but detailed studies of the nexus between export agriculture and poverty/food insecurity alleviation frequently belie the simplifications inherent in the proponents of liberalization (see Berry (2001) for a good outline of these arguments).

In summary, the ideological privileging within the WTO for resolving food security via the invisible hand of the market has the effect of generating policy myopia with regards to deeper and more nuanced understandings of the grounded relationships between trade policy and food security. This does not deny the undoubted propensity for liberal markets to advance resource efficiency; as Watkins (2008, p155) attests, 'Under the right conditions, agricultural trade could act as a powerful force for poverty reduction'. However, the pursuit of these policies in contexts where key agreements are not framed around sophisticated understandings of the connections between trade and food security could exacerbate, rather than resolve, current problems. Member states at the WTO carry with them these understandings, which foster reluctance to endorse the liberalization project for agriculture. To a considerable extent, the collapse of Doha Round negotiations in 2008 (and their failure to be restarted in the period since) can be attributed to these factors.

## Conclusion

This chapter has focused on the disputed terrain of how trade and food security relate to one another. Its opening observation was that this is a fractured field, within which trade can be portrayed either as the answer or the obstacle to improved food security. Essentialist renderings of this relationship, however, do little to advance consideration of this issue. Clearly, the trade of agricultural and food products across international borders can play a crucial role in advancing global food security. In dispute is the issue of when and how this occurs. At the international level, however, institutional arrangements do not encourage the incorporation of such detail within the spaces of negotiation. Instead, the WTO's approach to the issue is coloured by a set of overarching ideological preferences, which have the effect of instinctively generating scepticism of debate on food security as 'back door protectionism'. As a result, entitlements-based and rights-based approaches to the issue receive minimal attention within the organization.

Perhaps it is timely to invoke a paradigm shift in the ways that trade and food security are linked at the international level. With the ongoing lack of traction in the Doha Round speaking to a foundering of the WTO system, 'a new world may be possible'. Recently, scholars in international law have given heightened attention to the concept of 'regime interaction' – the question of how to generate coherence between international systems with different origins and foci (Young, 2009; Underdal and Young, 2004). To date, much of this scholarship has concerned the interactivity between the WTO, international environmental institutions (such as the Convention on Biological Diversity) and putative agendas to craft international climate change agreements. It is not implausible, however, to imagine a scenario where food security concerns – based around entitlements-based and rights-based approaches – are mainstreamed into a future system of international trade law. Such incorporation would provide a new arena for members to legitimately temper liberalization based on the food security implications of market opening (something like a 'food security box', to use WTO-speak). This approach does not reject the overall potential for trade to improve food security, but contextualizes policy change in terms of its grounded implications for vulnerable populations.

In brief, the millions of currently food-insecure people on the planet deserve a better deal from the world community. It is not appropriate to operate the rules of global agricultural and food trade in such a way as to treat food security as a residual outcome deriving from assumed market processes. A new system of trade rules could – and should – link the post-1990s understandings of food security to the global management of trade. Then, and only then, could we be better assured that trade would work for the hungry.

# References

Berry, A. (2001) 'When Do Agricultural Exports Help the Rural Poor? A Political-Economy Approach', *Oxford Development Studies*, vol. 29, no 2, pp125–144

Chung, K., Haddad, L., Ramakrishna, J. and Riely, F. (1997) *Identifying the Food Insecure: The Application of Mixed-Method Approaches in India*, IFPRI, Washington, DC

FAO (1996) 'Declaration on world food security'. *World Food Summit*, FAO, Rome, Italy

FAO (2003) *Trade Reforms and Food Security: Conceptualising the Linkages*, FAOI, Rome, Italy, available at www.fao.org/DOCREP/005/Y4632E/y4632e06.htm#bm06

FAO (2004) *Voluntary Guidelines to Support the Progressive Realization of the Right to Adequate Food in the Context of National Food Security*, FAO, Rome, Italy

FAO (2006) *Food Security: Policy Brief*, available at: ftp://ftp.fao.org/es/ESA/policy-briefs/pb_02.pdf

Fold, N. and Pritchard, B. (2005) (eds) *Cross-continental Food Chains*, Routledge, London

Friedmann, H. (1994) 'Distance and Durability: Shaky Foundations of the World Food Economy', in P. McMichael (ed) *The Global Restructuring of Agro-Food Systems*, Cornell University Press, Ithaca, pp258–276

Le Heron, R. (1993) *Globalized Agriculture: Political Choice*, Pergamon, Oxford

Levendal, C. R., Knowles, M. & Horii, N. (2004) 'Understanding vulnerability to food insecurity: Lessons from vulnerable group profiling', *ESA Working Paper* 04–18, FAO, Rome, Italy

OECD (2010) *Producer and Consumer Support Estimates*, OECD. Database 1986–2009, online resource, Database 1986–2009, online resource, www.oecd.org/tad/support/psecse, accessed 10 December 2010

Oxfam (2005) 'WTO agreement a betrayal of development promises', Media Release, www.oxfam.org/en/news/pressreleases2005/pr051218_wto_betrayal, accessed 18 December 2010

Potter, C. (2006) 'Competing narratives for the future of European agriculture: the agri-environmental consequences of neoliberalization in the context of the Doha Round', *The Geographical Journal*, vol 172, no 3, pp190–196

Pritchard, B. (2009) 'The long hangover from the second food regime: A world-historical interpretation of the collapse of the WTO Doha Round', *Agriculture and Human Values,* vol 26, pp297–307

Sen, A. (1981) *Poverty and Famines: An Essay on Entitlement and Deprivation,* Clarendon Press, Oxford

Underdal, A. and Young, O. (eds) (2004) *Regime Consequences: Methodological Challenges and Research Strategies*, Kluwer, Dordrecht, Netherlands

Watkins, K. (2008) 'Agricultural trade, globalization and the rural poor', in von Braun, J. & Diaz-Bonilla, E. (eds) *Globalization of Food and Agriculture and the Poor*, Oxford University Press and International Food Policy Research Institute, Oxford, pp155–180

WTO (2005) 'Why trade matters for improving food security', DG Speeches, *High-level Roundtable on Agricultural Trade Reform and Food Security*, Rome, www.wto.org/english/news_e/spsp_e/spsp37_e.htm, accessed 13 April 2010

WTO (2010) 'Glossary – food security' www.wto.org/english/thewto_e/glossary_e/food_security_e.htm

Young, M. (2009) 'Fragmentation or interaction: the WTO, fisheries subsidies, and international law', *World Trade Review*, vol 8, no 4, pp477–515

# 5

# BIOFUELS AND THE FINANCIALIZATION OF THE GLOBAL FOOD SYSTEM

*Philip McMichael*

## Introduction[1]

This chapter suggests the biofuels question is more complex than simply a zero-sum relation between fuel and food crops. While public authorities and development agencies puzzle over whether, and to what extent, fuel-cropping is a sensible and/or sustainable use of arable land, the deeper question concerns the implications of financialization of the global food system. As agriculture and food are increasingly incorporated into financial chains, a crop-mix calculus becomes ever more irrelevant and inapplicable, as decisions over material production are driven by an abstract financial calculus in corporate boardrooms with little concern for food security and/or environmental integrity.[2]

The recent expansion of industrial biofuels expresses several trends in global political-economy, including the global commodification of a time-honoured local energy supplement, and the consolidation of corporate power in the energy and agribusiness sectors. Manifestly, the 'biofuels revolution' is about the energy crisis, as the cost of capital inputs (production, processing, transport) rises in an age of peaking oil supplies. In addition, reduction of dependence on Middle Eastern oil drives governments around the world to develop an industrial biofuels complex to relieve energy insecurity. At the same time, biofuels represent a new profitability frontier for agribusiness and energy sectors beset with declining productivity and/or rising costs. Questions about reducing or transforming energy-use patterns have been tabled while politicians look to short-term job creation to recover from the deflating effects of a financial crisis – so while economic growth remains inviolate, biofuels, as 'green energy', are represented as a solution to environmental concerns. Simultaneously, the resort to biofuels symbolizes the fetish of exchange-value as, on the one hand, fuel crops displace food crops and, on the other, deepen a process of global agro-industrialization premised on land-grabbing, displacement of small producers, and increased emission of carbon (deforestation) and nitrous oxide (inorganic fertilizer). The privileging of exchange-value over use-value ('ecological capital' including farming knowledge) subordinates agriculture to a financial calculus at the expense of socioecological sustainability.

As such, the biofuels revolution condenses some critical thresholds in global political-economy. These include, first, the substitution of fuel crops for food crops – signalled by critics renaming biofuels 'agrofuels'; second, the 'financial turn', as speculative investments drive the development of an agrofuels project; third, the global land-grab, as governments conspire with development agencies and investors to enclose land for food and energy plantations; and fourth, the consolidation of a process of capital involution, as the expanded reproduction of capital depends on the centralization of agro-industrial capital via cycles of social exclusion. The latter process, normalized as 'development', is clearly represented in the agrofuels project – the subject of this chapter.

## The agrofuels project

The recent emergence of *industrial* biofuels, with subsidies and cross-sectoral and speculative investment, amounts to a coordinated 'agrofuels project' (McMichael, 2009b), targeting land in the global South. In 2007, the US government set corn ethanol targets (35 billion gallons by 2017) with huge subsidies to the agribusiness giants such as ADM, Bunge, and Cargill. The European Union matched this with a 10 per cent target by 2020 for a biofuels mix in transport fuels. The UK Gallagher Report (2008) estimated that by 2020 about 500 million more hectares of land, one-third more than currently under cultivation, would be required to meet global demand for biofuels. To accomplish this, not only would 70 per cent of European farmland need conversion to fuel crops, but also the entire US corn and soy harvest (Holt-Giménez, 2007a). However, arable land prices in the US rose 13 per cent in 2007, and over 10.5 per cent in 2008; while in the UK prices rose 28 per cent in late 2007, and by more than 10 per cent in the first quarter of 2008 (Berthelot, 2009, p16). And so, given (subsidized) biofuel targets *and* enabling Kyoto protocols, corporations and financiers are increasingly investing in agrofuels, production offshore – as depicted in Table 5.1, recording venture capital investments in biofuels for 2009.

European firms already claim over five million hectares of Southern land for biofuel development, and Southern governments accommodate such 'agro-imperialism' in the name of debt repayment (via export revenues and rural development), and in the service of local patronage networks. Recent plans are for Brazil to replace 10 per cent of the world's fossil fuels by 2025 with sugar ethanol; Malaysia and Indonesia are expanding oil palm plantations to supply 20 per cent of EU bio-diesel needs; India plans 14 million hectares of land for jatropha plantations, and Africa 400 million (Holt-Giménez, 2007a; Vidal, 2007, p3).

Such a biofuels complex deepens exploitation of Southern resources (land, water, labour combined with industrial fertilizers and pesticides) to supplement fuel consumption by high-energy consumers with cheaper forms of ethanol and biodiesel, and without reducing the total amount of emissions. In the name of rural

*Table 5.1* Biofuel investments by country, 2009

| Country | Investment (in US$) |
| --- | --- |
| Brazil | 3.45 billion |
| US | 2.01 billion |
| Papua New Guinea | 800 million |
| Canada | 589 million |
| India | 438 million |
| Rwanda | 250 million |
| Philippines | 197 million |
| South Africa | 190 million |
| Ethiopia | 84 million |
| Australia | 75.3 million |
| Sweden | 73 million |
| Kenya | 46 million |
| Argentina | 37 million |
| South Korea | 17.6 million |
| Peru | 12 million |
| Mozambique | 10 million |
| Poland | 4.3 million |

Source: *Biofuels Digest*, www.biofuelsdigest.com

development – and foreign-debt repayment – indebted states embrace biofuel (and food) investments, facilitating a 'global land-grab' that compromises previous land reform commitments made, for example, at the 2006 International Conference on Agrarian Reform and Rural Development (de Schutter, 2010).

The agrofuels project is symptomatic of the phenomenon of financialization,[3] whereby investors prefer to hold capital in liquid (rather than illiquid/asset) forms. Arrighi (1994) has argued that 'financialization' signals declining hegemony, as a lead state's productive capacity loses its competitive edge in international political-economy, and its industrial capitalists switch investment from fixed capital into financial channels. This liquidity preference has encouraged securitization (consolidating and selling debt), mergers (including firm acquisition by private equity companies that unbundle unprofitable units for financial gain) and general financial speculation.[4]

Financialization has been associated with a global decline in productivity outside of the information and communications technology sectors, as financial gains have been more attractive – particularly during an era in which production has relocated to Southern regions which provide cheap land and labour through subcontracting systems. Meanwhile, for a time, Northern consumption of such offshore products was sustained by the banking revolution, involving profligate mortgage lending and rising consumer debt. By the twenty-first century, however, the decline in industrial productivity, combined with the collapse of the financial derivatives market, resulted in an accumulation crisis. One notable consequence has been the decisive shift of investment capital into speculative ventures in land, food and biofuels. For example, between 2004–07, venture capital investment in biofuels increased by 800 per cent (Holt-Giménez, 2007, p10).

## The agricultural investment frontier

The question here is: why has agriculture become a significant frontier of investment? There are two – related – explanations. One is that food itself became a speculative investment through the device of commodity index funds. In this scenario, 'agrofutures' became targets of investment (alongside energy and industrial metals)[5] as agricultural contracts were converted into derivatives, following pressure in the 1990s, by financiers on legislators, to deregulate the commodity contract business. A market in food became a market in food contracts, taking on a life of its own (counting on rising derivative prices, as such futures were traded repeatedly). What was once a mechanism of hedging risks on food prices for producers and consumers, reducing volatility, became a virtual mechanism whereby financiers constructed commodity index funds with which they could profit from price volatility without risk – by harbouring large percentages of clients' index fund investments in safer ventures and then profiting from rising, or declining, food prices (Kaufman, 2010, pp30–31). Buying and selling food futures, then, developed into a derivative market, which in turn inflated food prices. Such speculation was enabled by automation in computing, and intensified during the first decade of the new century as the real estate market crisis unfolded. At that point, investors shifted funds into commodity futures: between 2003 and 2008 commodity index holdings increased from $13 billion to $317 billion (Kaufman, 2010, p32). This speculative spike resulted from the process whereby 'the mechanism created to stabilize grain prices had been reassembled into a mechanism to inflate grain prices' (Kaufman, 2010, p34).[6]

The other explanation is that, in the context of an accumulation crisis and the financial crisis (symbolized in the sub-prime mortgage shake-out of 2008–09), international capital markets have gravitated towards agriculture as a relatively safe investment haven for the relatively long-term. Most notably, in 2007, 'soft' commodities (renewable crops) overtook 'hard' commodities (non-renewables, such as oil and metals) as 'prime performers' in the commodities investment market. Researchers have cited new demands for bioenergy and other 'bioproducts' from agricultural crops among the causes of this 'bull-run on soft commodities' (Daniel, 2009, p5). Some examples of this trend in agricultural financialization are presented in the box overleaf.

These comments clearly signal an important threshold in the history of the 'food regime': a concept underlining the political role of agricultural commodification and circuits of food in the institutional development of world capitalist economy (Friedmann and McMichael, 1989). In this instance, the threshold represents a

## Forms of agricultural financialization

In June, 2009, the Executive Director of JP Morgan, stated: 'Physical agriculture's assets are the new focus in longer term investments as institutional investors explore opportunities in everything from raw land to grain elevators to food processing plant' (quoted in Gillam, 2009).

In Southeast Asia: 'PT Daewoo Logistics Indonesia, a subsidiary of South Korea's Daewoo Logistics Corporation, and Cheil Jedang Samsung recently announced a partnership to invest $50 million to grow and process energy crops on the Indonesian islands of Buru and Samba ... [for export] back to South Korea. In early 2008, Sinopec and the Chinese National Overseas Oil Corporation, two state-owned oil giants, made investments in Indonesia of $5 billion and $5.5 billion, respectively, to grow and process corn into biofuel to be exported to China' (Daniel, 2009, p4).

On June 5, 2008, *The New York Times* published a lead article, 'Food is Gold, So Billions Invested in Farming', with some key statements, as follows:

- 'It's going on big time. There is considerable interest in what we call 'owning structure' – like United States farmland, Argentine farmland, English farmland – wherever the profit picture is improving' (Brad Cole, president of Cole Partners Asset Management in Chicago – managing a hedge fund for natural resources).
- 'The world is asking for more food, more energy. You see a huge demand ... What this new investment will buy is new technology. We will be helping to accelerate the development of infrastructure, and the consumer will benefit because there will be more supply' (Axel Hinsch, CEO of Calyx Agro, a division of Louis Dreyfus Commodities, owning tens of thousands of acres of cropland in Brazil, backed by large institutional investors, like AIG).
- Susan Payne, founder and CEO of Emergent Asset Management (UK), which is raising $450–750 million for investment to consolidate and industrialize farmland in sub-Saharan Africa, for food and biofuels (jatropha): 'We are getting strong response from institutional investors – pensions, insurance companies, endowments, some sovereign wealth funds.' The fund selected Africa because 'land values are very, very inexpensive, compared to other agriculture-based economies. Its microclimates are enticing, allowing a range of different crops. There's accessible labour. And there's good logistics – wide open roads, good truck transport, sea transport.'

significant deepening of agro-industrialization in a crisis conjuncture.[7] Hitherto, manufacturing and services have been the principal targets of Northern investment. Both of those sectors have been outsourced and offshored during the past quarter of a century, in a spatialized process of labour cheapening and labour disorganization. While agro-industrialization has concentrated in the global North, producing both a cheap food regime as a complement to Northern manufacturing dominance (McMichael, 2005) and a weapon to dispossess Southern peasants (sustaining consumerism via exploitation of a seemingly endless reservoir of cheap prematurely-urbanized offshore labour), it now appears to have reached its limits of profitability. The solution is not simply an inter-sectoral shift of capital (from manufacturing to agriculture); rather it is replication of a similar pattern to offshore manufacturing: namely, relocating to exploit cheap inputs.

The profitability crisis for industrial agriculture is two-sided: it refers to the declining biophysical productivity of industrial agriculture – for example, the drop in efficiency of nitrogen use from 60 to 20 per cent from the 1950s to the 1990s (Van der Ploeg, 2010, p100) and loss of biodiversity and crucial 'ecosystem services' such as pollination and soil formation (Weis, 2010, p316); and to the attempts at 'biophysical override', as capital compensates by deepening the commodification of production (Weis, 2007). Both trends drive agro-capital offshore to appropriate Southern lands. This relocation of investment is compounded by anticipation of risk evident in climate change.[8]

While much has been made of the productivity gains associated with agro-industrialization via the green revolution, for example, it has evidently run its course as its 'external' costs rise – notably pest resistance, soil erosion, water table depletion and farmer debt (Sharma, 2004; Zwerdling, 2009), and rising use of energy and irrigation water with declining efficiency of use (van der Ploeg, 2010, p100). That revolution was initially profitable for upstream supply firms (seeds, machinery, agro-chemicals). However, the upstream and downstream companies servicing the grain corporations, accounting for 20 per cent of food expenditures, have experienced a profit squeeze as rising investments (chemical inputs, genetic engineering and mechanization) have not been matched by rising productivity rates (Holt-Giménez. 2007b, p10). The consequence has been the development of two, related tendencies: (1) the centralization of capital into 'financialized' food conglomerates (Rama, 2005; Burch and Lawrence, 2009) to manage the profit crunch; and (2) rising offshore investment in agro-industry in the global South, where land is cheap and labour plentiful.

These tendencies express contemporary world-market ordering of agricultural production and circulation, conceptualized as 'food empires' which 'increasingly exert a monopolistic power over the entire food supply chain' (van der Ploeg, 2010, p99). Industrialization and 'globalization' processes require 'permanent scale increases', triggering 'a heightened dependency of farming upon capital markets', increasing debt and putting a premium on short-term financial gains (van der Ploeg, 2010, p100). The result is a 'world agriculture' (McMichael, 2005) or 'the complete interchangeability of large agricultural systems' (van der Ploeg,

2010, p101) where high-value industrial crops are grown anywhere, expressed in the export of food processing and supermarkets to colonize Southern urban consumption patterns (Reardon and Timmer, 2005). Meanwhile, the financialization of food empires (for example Parmalat, Ahold, Carrefour, Charoen Pokphand, Salim Group, Fonterra) has exerted increasing pressure on food prices at each end of the supply chain: namely, rising for consumer foods, and declining for farmer produce (Rosset, 2008; McMichael, 2005). Van der Ploeg (2010, p102) points out that in both Italy and the Netherlands during the last two decades of the twentieth century, total value-added indexes for industrial subsectors (for instance chemicals, wood, auto, mineral processing, textiles) rose considerably more slowly than those for the food industry, underlining the attraction of agro-food investments for financial capital.

## Agriculture as capital's new frontier?

The agro-industrialization of the global South is a logical response to capital's crisis of profitability. Rising food prices, peaking oil, cheap land, and stalled investment funds together drive a new solution to the accumulation crisis, represented as serving the interest of humanity and the planet. This would appear to be a feasible explanation for shifting financial flows into agriculture, specifically in the global South, where environmental regulations are weak or non-existent. Whether 'agriculture' itself is the answer to the accumulation crisis is not necessarily the point, given Northern subsidies to agribusiness, energy and transport companies, and Southern concessions to investors.

Noting that agriculture is generally considered to be an inferior arena for added value, François Houtart asks: 'How, therefore, can agriculture become a new frontier for the accumulation of capital?' (2010, p127). Because of the limits on realization of agricultural capital given by food demand inelasticity, compounded by the active marginalization of the majority of humanity (as consumers), the only feasible answer for Houtart is agrofuels, 'which have come just in time to revive the prices of agricultural products and their role as a financial refuge in times of crisis' (2010, p128).[9] Affirming this view, the UN reported in 2007 that biofuels were the fastest growing segment of the world agricultural market (ETC, 2007, p2), fuelled by cross-sectoral (and infrastructural) alliances between energy, agribusiness, trading companies, hedge funds, sovereign funds, states, UN agencies and universities.[10] The expansion of an agrofuels frontier in the global South represents an appropriation of cheap (or free) land and low-cost labour on industrial plantations of sugar cane, soy, eucalyptus, oil palm, maize, wheat and jatropha.

The cost of land may be low for investors, but not to the locals.[11] For example, in Ethiopia (one attractive investment site for agrofuels), an agriculture ministry official identified over seven million acres of 'virgin land', to be leased at an annual rate of about 50 cents per acre (Rice, 2009). Ethiopia's 'land lease project', intended to develop large-scale commercial farming (mainly for export of food and fuel), will involve the allocation of three million hectares of 'idle land' by 2013 (about 20 per cent of currently cultivated land).[12] However, an indigenous Anuak from the fertile Gambella region of Ethiopia observed: 'All the land round my family village of Illia has been taken over and is being cleared. People now have to work for an Indian company. Their land has been compulsorily taken and they have been given no compensation' (quoted in Daniel, 2010, p28). International Land Coalition policy specialist, Michael Taylor, has noted:

> If land in Africa hasn't been planted, it's probably for a reason. Maybe it's used to graze livestock or deliberately left fallow to prevent nutrient depletion and erosion. Anybody who has seen these areas identified as unused understands that there is no land in Ethiopia that has no owners and users.
>
> quoted in Daniel (2010, p20)

In other words, cheap land is available as a subsidy to investors by governments trading away social reproduction rights of peasants. Essentially this amounts to an ongoing process of 'primitive accumulation', whereby capital's profitability depends on *publicly-managed expropriation*. Agrofuels may be the new frontier of capital accumulation, but they depend on subsidies from home and host governments. A recent World Bank study 'found that large-scale export agriculture in Africa has succeeded only with plantation crops like sugar and tea or in ventures that were propped up by extreme government subsidies, during colonialism or during the apartheid era in South Africa' (Rice, 2009); it is not difficult to extrapolate forward to current conditions, where, according to Friends of the Earth and EarthTrack, the combination of the Renewable Fuels Standard Mandate (which provides a market for biofuels) with tax credits would subsidize the US biofuels industry to the tune of $400 billion through to 2022.[13] Analyst Bloomberg New Energy Finance reported that 'in 2009 governments provided subsidies worth between $43bn (£27bn) and $46bn to renewable energy and biofuel industries, including support provided through feed-in tariffs, renewable energy credits, tax credits, cash grants and other direct subsidies' (Business Green, 2010).

Subsidization of industrial biofuels underpins the 'agrofuels project', based as it is on the 'externalization' of a number of 'costs'. These costs include the rights of small farmers to ancestral lands; food insecurity arising from the conversion of food-producing land to food- or fuel-crop export agriculture; environmental deterioration resulting from industrial agriculture; and increased greenhouse gas emissions. Each of these issues eventually become monetary (and opportunity) costs, as governments and development agencies are confronted with displacement, food shortages, and ecological disruptions.

In other words, the 'biofuels revolution' is not simply about restoring profitability to capital via investment fund management within a subsidy regime. It is also about pursuing an agro-imperial development trajectory premised on sacrifice – of land and its inhabitants to a financial calculus. In what follows, I argue that the commercial development trajectory embodied in the agrofuels project authorizes large-scale removal of peasant populations from ancestral lands to install 'agriculture without farmers' – as the international peasant coalition, La Vía Campesina, calls agro-industrialization. In this process, the agrofuels project follows a diabolical logic: arresting a crisis of profitability for capital, at the expense of human and natural ecology.

## The land-grab complex

The 'global land-grab'[14] combines the domestic construction of land rents with *new* mercantilist food security practices, as foreign governments sponsor offshore agriculture to ensure national food and energy security.[15] Assisted by World Bank policy, the land-grab is represented as a form of development, insofar as land 'development' is associated with productivity gains and employment, and indebted governments in the global South stand to receive foreign investment and hard currency from conversion of their land and forests into agro-export platforms. Biofuels in particular claim a new role in development. Thus, President Andrés Pastrana of Colombia sought, in 2001, to attract Malaysian investment in a three million hectare oil palm project by claiming: 'progress and social development can reach large areas of Colombia that are ready to join in the cultivation and processing of this primary commodity' (quoted in Escobar, 2008, p85).

The NGO (non-governmental organization) sector follows a similar logic, arguing that biofuels generate employment through rural diversification. Oxfam (2007, p5) states in its 'Biofuelling Poverty' report that:

> Biofuels need not spell disaster for poor people in the South – they should instead offer new market and livelihood opportunities. But the agro-industrial model that is emerging to supply the EU target poses little in the way of opportunities and much in the way of threats.

Its solution, however, is to propose a set of social principles governing the development of a biofuels industry. Complementing Oxfam's social vision is the UK Gallagher Report (2008), which cautions against displacing food crops, but nevertheless suggests that alternative energy crops can simultaneously provide new employment and local development opportunities to rural communities. By contrast, estimates are that in tropical regions:

> 100 hectares dedicated to family farming generates 35 jobs. Oil-palm and sugarcane provide ten jobs, eucalyptus two, and soybeans a scant half-job per 100 hectares, all poorly paid ... Hundreds of thousands [of smallholders] have already been displaced by the soybean plantations in the "Republic of Soy", a 50m hectare area in southern Brazil, northern Argentina, Paraguay, and eastern Bolivia.
>
> Holt-Giménez (2007a, p10)[16]

The enclosure of land for agrofuels, and food, in the global South is revitalizing a long-standing (but institutionally dormant) modernization trope, namely that modernization of agriculture is necessary to development. The recent World Bank *World Development Report* (2008),[17] centred on 'agriculture for development', was the first time in a quarter of a century that this key development institution paid attention to agriculture. It appears that the urgency of the food and energy crises has refocused the attention of the global political-economic elite on mobilizing agricultural resources to offset food, water and fuel shortages. Agricultural land in the global South, in particular, is targeted for 'productivity increase' via technification. For example, Susan Payne, CEO of Emergent Asset Management (a UK investment fund planning to spend $50m on African land) declared:

> Farmland in sub-Saharan Africa is giving 25 per cent returns a year and new technology can treble crop yields in short time frames ... Agricultural development is not only sustainable, it is our future. If we do not pay great care and attention now to increase food production by over 50 per cent before 2050, we will face serious food shortages globally.
>
> quoted in Vidal (2010)

In Africa, much of the land is communally held, but is subject to government designation as 'idle' land,[18] and given potential rewards through commercialization. Unsurprisingly, international development and financial institutions are working behind the scenes on privatising land relations to enable and attract foreign investment in African land. US investment, for example, is encouraged by the US government's Millennium Challenge Corporation (MCC), which disburses money in the form of grants to particular countries on condition that they meet certain neoliberal economic criteria. Most MCC Compacts signed with African countries focus on agriculture, with a central land privatization component supporting 'market-based solutions to food security'. Such provisions include certifying outgrowers for food exports; constructing infrastructure to gain access to world markets; and partnering with The Alliance for a Green Revolution in Africa (AGRA) to provide inputs to farmers in their first year (GRAIN, 2010a).

Infrastructure is critical to the land-grab and includes a complex of processes of legal adjustments to privatize land; new codes of conduct (Borras and Franco, 2010); public-private partnerships to build biofuel production and refining capacity;[19] certification schemes to standardize biofuel products for the world market; and justifications that intersect with Kyoto's Clean Development Mechanism protocols – encouraging financiers from the North (where emission targets and caps are in force) to invest in green projects in the global South, where emission caps are non-existent. An exemplary case of the latter is the intention of the US transnational corporation Arcadia Biosciences to finance the development of GM crops in China, starting with rice, and following with wheat, maize, sugar cane and oleaginous plants – the patent for which Arcadia has already sold to several companies, including Monsanto. As Houtart (2010, p687) explains, this is planned 'through the compensations envisaged by the carbon trading market. The reasoning is that, with the GM seeds supplied by Arcadia, peasants will use much less fertilizer that emits nitric oxide. As this is 300 times more harmful than $CO_2$ the peasants could finance the purchase of the GM seeds'. Houtart (2010, pp69–70) notes that infrastructure and rationale for such agro-industrialization is provided by the Global Leadership for Climate Action group (Club of Madrid), dedicated to public-private initiatives to reinforce democracy and address global challenges, while 'socializing the risks and privatizing the profits'.

The global land-grab is promoted by, among other organizations, the World Bank; its International Finance Corporation (IFC);[20] the International Rice Research Institute (IRRI) of the Consultative Group on International Agricultural Research (CGIAR); the European Bank for Reconstruction and Development (IBRD); and others, with particular focus on sub-Saharan Africa. The IFC, for example, formed an alliance in early 2009 'with Altima Partners to invest in farming operations and agricultural land in "emerging market countries"'. The new $625 million Altima One World Agricultural Development Fund is IFC's largest equity investment in its expanding agribusiness portfolio' (Daniel 2009, p6). The IFC's partner, The Foreign Investment Advisory Service (FIAS), targets 'investment climates' in foreign markets, creating land registries and easing the process of land titling, leasing and foreign investment – made easier where, as the International Institute for Environment and Development (IIED) found, 'many countries do not have sufficient mechanisms to protect local rights and take account of local interests, livelihoods, and welfare' (Houtart 2010, p17).

In 2009, a new FIAS initiative – Investing Across Borders (IAB) – conducted project surveys in 87 countries, targeting information regarding technical regulatory and licensing only. However, it discounts human impact: 'nothing about the IAB indicators seeks to consider the extent to which local populations in these countries will be affected – whether local populations already occupy the land, whether the land provides water supply or grazing lands for local populations, etc.' (Daniel, 2010, p15). Furthermore, the information compares investment climates and opportunities for competitive purposes. Thus the IFC/FIAS compiled 'A Diagnostic Checklist for Land Markets' that itemizes questions about land-holding customs, law, power struggles, state capacity to protect investments and so on, in addition to a Benchmarking FDI (Foreign Direct Investment) Competitiveness Report (2007), noting for Kenya, for example, 'strengths' in the horticulture sector such as 'abundance of arable land', 'low employment rigidity', 'low air transit costs for shipments to Amsterdam', and so on. For Tanzania, it pointed out that '(t)he country has an abundance of arable land [only 5.5 per cent of which is utilized] from which horticultural operations can be established' (quoted in Daniel, 2010, pp17–18). Such development 'services' constitute a broad infrastructural complex supporting land-grabbing – both material and ideological.

A GRAIN report, relied on by the World Bank in its report on large-scale land acquisitions,[21] draws on 389 land deals in 80 countries, where the 'bulk (37 per cent) of the so-called investment projects are meant to produce food (crops and livestock), while biofuels come in second place (35 per cent). Unsurprisingly, Africa is the target of half the land-grab projects, followed by Asia, Latin America and Eastern Europe' (GRAIN, 2010a). GRAIN's report notes that European Investors claim protection from the Bank's Multilateral Investment Guarantee Agency for 'political risk insurance for their farmland deals' (GRAIN, 2010a).

## Land-grabbing's developmentalist assumptions

The Gates Foundation suggests that enabling the commercial development of African agriculture 'will require some degree of *land mobility* and a lower percentage of total employment involved in direct agricultural production' – a clear allusion to eviction (Xcroc, 2009). According to GRAIN (2008), over US$100 billion has been mobilized since the food crisis summit in Rome in June 2008 for land invest-ments 'not to harvest food but to harvest money'. Given this rush to invest in land, questions of legitimacy have arisen, addressed in part by the enunciation in 2010 of seven 'Principles for Responsible Agricultural Investment' by the World Bank; the FAO; its International Fund for Agricultural Development (IFAD); and the United Nations Conference on Trade and Development (UNCTAD) Secretariat. While these principles claim to benefit investors and affected communities alike, they nevertheless provide an unequal comparative advantage to investors, given the relationship between privatising land and its 'mobility'.[22]

IFC Performance Standards on Social and Environmental Sustainability are quite explicit concerning displacement of rural peoples. For example, Performance Standard #5: Land Acquisition and Involuntary Resettlement refers 'both to phys-ical displacement (relocation or loss of shelter) and to economic displacement (loss of assets or access to assets that leads to loss of income sources or means of livelihood) as a result of project-related land acquisition' (quoted in Daniel, 2010, p49). The text continues:

> Unless properly managed, involuntary resettlement may result in long-term hardship and impoverishment for affected persons and communities, as well as environmental damage and social stress in areas to which they have been displaced. For these reasons, involuntary resettlement should be avoided or at least minimized. However, where it is unavoidable, appropriate meas-ures to mitigate adverse impacts on displaced persons and host communi-ties should be carefully planned and implemented. Experience demonstrates that the direct involvement of the client in resettlement activities can result in cost-effective, efficient, and timely implementation of those activities ... Negotiated settlements help avoid expropriation and eliminate the need to use governmental authority to remove people forcibly. Negotiated settle-ments can usually be achieved by providing fair and appropriate compensa-tion and other incentives or benefits to affected persons or communities, and by mitigating the risks of asymmetry of information and bargaining power. Clients are encouraged to acquire land rights through negotiated settlements

wherever possible, even if they have the legal means to gain access to the land without the seller's consent.

Beyond the assumption that some, and eventually more, communities will be displaced into resettlement areas 'with appropriate disclosure of information, consultation, and the informed participation of those affected', there is the additional assumption, spelt out in Performance Standard 7, that 'Private sector projects may create opportunities for indigenous peoples to participate in, and benefit from, project-related activities that may help them fulfill their aspiration for economic and social development' (Daniel, 2010, p50).

Basically, eviction is to be handled as an inevitable price, or opportunity of progress, understood as agro-industrialization to fuel and energize the global consumer class.

Representation of the land-grab by development agencies as a 'win-win' operation proceeds from neo-classical assumptions that development necessitates rural depopulation.[23] Haroon Akram-Lodhi (2008) notes the Bank's *World Development Report, 2008*, 'clearly expects that over time agriculture-based countries should, eventually, shift to becoming transforming countries before, eventually, becoming urbanized countries'. This large-scale transition is premised on developmentalist assumptions about agro-industrialization, ultimately rooted in labour displacement and land concentration.[24] Viewing agriculture as the servant of growth, the emphasis is on plant yield – rather than agricultural function – in order to develop capacity for seeds and fertilizer inputs. An alternative conception of raising productivity is as 'food output per acre rather than yield per plant' where farming is diverse, needing investment in 'ecologically sound and socially just technologies' (Murphy, 2008). This trajectory, however, is at odds with the appropriation of farmer knowledge as the condition of agribusiness centralization, and, therefore, 'development'.[25]

De-peasantization is the ultimate development litmus test, including the enclosure of smallholders within what the World Bank calls the 'new agriculture', 'led by private entrepreneurs in extensive value chains linking producers to consumers and including many entrepreneurial smallholders supported by their organizations' (World Bank, 2007, p8). The artificiality of this conception is that it is *synchronic*, advocating instant incorporation of smallholders into a hierarchical global market structure, rather than the assumed 'evolutionary' (diachronic) process within particular countries. Amin (2003) has reclassified this agrarian hierarchy as: high-input grain-livestock farmers in the North; a relatively small group of industrial-capitalist farmers in the agro-industrial export regions of the South; and the globally pervasive, underprivileged, low-input smallholder population (which comprises about 40 per cent of humanity (Araghi, 2000)). Within this hierarchy, agricultural productivity ratios across high- and low-input farming have risen from 10:1 prior to 1940 to 2000:1 in the twenty-first century, deepening the competitive exposure of smallholders (Amin, 2003, p2). By conflating a diachronic evolutionary assumption with a synchronic regime – whose competitive advantages reside in subsidized agri-business trade and investment at the expense of peasant agriculture – the World Bank consigns peasants to a residual status, thus implying their justifiable replacement in the name of human progress.

The evident obsolescence of smallholder agriculture in the context of a market-driven project of managed competition, with complicity by government and development agencies in land-grabbing, rests on a development episteme unable to recognise the social, ecological and cultural functions and potentials of small farming practices and networks. Liberal NGOs such as Oxfam also consider investing in agriculture as likely to have 'an enormous poverty reduction "pay off", because of agriculture's importance to food security …'. Thus the Oxfam International Research Report: *Harnessing Agriculture for Development* claims 'Agriculture is certainly an important part of the mix of activities that sustain household economies, but has to be viewed in the context of increased multi-activity by poor households, deepening urban-rural linkages and heightened national and international out-migration'. Acknowledging that 'certain features of small farms – their transmission of local knowledge for instance – can also mean they have a key role to play in protecting environmental goods', the executive summary continues 'it may be necessary to recognize that, in some cases, invest-ment in agriculture will be about enabling rural populations to exercise greater choice about their livelihoods, including leaving farming altogether' (quoted in Fraser, 2009). Ultimately, then, peasant agriculture is regarded as development's 'poverty baseline' for development,[26] and largely accepted as such in the official development paradigm.

Given that a key register for development is the (apparent) absence of peasant-ries in the global North,[27] policy-makers, funders and commentators organize their data along these lines, making the assumption that there is a singular (standard) trajectory in play, governed by scale efficiencies, market-rational resource allo-cation, and so forth. Peasant migration off the land is thus a function of either economic underachievement or simply choice.[28] Ultimately, the point is that the standard metric in play privileges a 'false economy', so to speak, constructed entirely of monetary measures informing the national accounting apparatus through which states conduct their business. One aspect of this concerns World Bank monitoring and evaluation, where, for example:

> FIAS indicators for project-specific "impact" include FDI/GDP statistics, gross fixed capital formation, export performance and/or private investments in specific industries, and the number of new business registrations. Nowhere within its M&E [monitoring and evaluation] does FIAS consider, for example, the number of local jobs created, changes to hunger and poverty statistics, the average incomes of local population, or whether [there is compliance] with IFC's own Performance Standards.

> Daniel (2010, p27)

Beyond recording its accomplishments one-sidedly (which reinforces perceptions of 'development', which in turn supports global business-as-usual), the other aspect of such metrics is that they externalize costs not reducible to monetary values (for instance human displacement, soil erosion, biodiversity loss, GHG (greenhouse gas) emissions), and discount other values such as socioecological sustainability, energy sovereignty, local food security and dietary diversity. Arguably, such one-dimensional celebration of rising development indices is symptomatic of deepening commodity fetishism, where price is a proxy for an entirely abstract notion of (exchange) value that actively obscures the concrete values by which humanity might live, and survive. The concluding section develops this point.

## Conclusion

The corporate food regime has progressively modelled a form of agriculture valuing its product solely as a commodity. That is, agriculture's use-value is conflated with exchange-value, rendering crops as fungible investments – as in the multiple uses of corn, soy, palm oil, and sugar, for example, whether as foods, feeds, fuels, cosmetics, stabilizers, and so on. For the crops mentioned, their conversion from food to exchange-value, magnified by financialization, perhaps represents the ultimate fetishization of agriculture as an input-output operation governed solely by profiting from indiscriminate production of commodities. In this context, fuel crop competition for food crop-land puts considerable pressure on world food prices, transmitted through 'food empire' supply chains.[29] In addition, the structuring of the corporate food regime has rendered global food security dependent on cheap agro-industrial surpluses traded globally (the US for 40 per cent of corn, along with Brazil and Argentina – who, as a threesome, account for 80 per cent of soybeans, with wheat being exported by the ex-settler states and a few EU countries). This, in a context where the volume of cereal imports into Low-Income Food-Deficit Countries (LIDFC) almost tripled between 1975–7 and 2005–07, with their prices rising 37 per cent in the three years before 2008 (Weis, 2010, p328).

In a further contribution to food price inflation, under the conditions of the agrofuels project, corporate hedging between fuel and food for these commodities decisively delinks agriculture from basic social reproduction.[30] Agro-industrialization is ultimately about combining commodified inputs (seeds, fertilizer, antibiotics, privately-owned genetic materials, pesticides and so on) with land or water or factory farms to produce outputs as ingredients of processed commodities to fuel labour or machinery without regard for ecological context. In other words, the process of abstraction is not simply about the destination of the produced commodities, but also about biophysical process. The conditions of fungibility are also premised on 'biophysical override' (Weis, 2007) – a condition of disregard for, or 'externalization' of, environmental consequences.

It is at this moment that the limits of agro-industrialization become not simply material (and social) but also epistemic. This is the unassailable interpretation of

UN Human Rights *rapporteur* Jean Ziegler's claim in 2007 that biofuels are a 'crime against humanity' — insofar as they undermine food-provisioning cultures, and degrade environments (increasing emissions,[31] with further degradation). In short, agrofuels fundamentally contradict the meaning of *social* reproduction. The epistemic point is that while agrofuels may be rational in terms of an investment portfolio (especially with massive public subsidies), they contradict planetary and human sustainability. Encouraged by an artificial 'market calculus', they are nevertheless socially, economically[32] and ecologically irrational. Agrofuels are not only water-intensive,[33] but they accelerate the export of water from the global South, embodied in agro-exports – exemplified by the participation of Middle Eastern countries in land-grabbing.

The social irrationality is that agrofuels increase emissions, cannot solve the energy crisis, and threaten existing common-lands, prairies and forests upon which a large portion (and ultimately all) of humanity depends. Of course, some smallholders gain access to additional income from leasing part of their land, or contracting with biofuel traders/processors (McMichael, 2010). But this is most likely a short-term phenomenon, ultimately entrapped in the development metric, reinforcing an unsustainable trajectory. What may be partial success in conventional rural development terms discounts the inability of agrofuels to solve the global energy crisis, let alone the crisis associated with marginalization of smallholders, pastoralists and forest-dwellers. At the same time, it discounts the fundamental importance of biodiversity preservation for planetary survival. Development entrapment means ignoring, discounting and marginalizing *extant* practices that have the potential to reduce energy profligacy, fossil-fuel dependence, and reorganize social life around such concrete and sustainable trajectories as food and energy sovereignty (Wittman, Desmarais and Wiebe, 2010).

A *global* biofuels market is still incipient (Wilkinson, 2009), as neo-mercantilist practices (protected, subsidized national biofuel sectors, with offshore complements managed through tariff structures) continue alongside emergent globalizing corporate/state arrangements – which anticipate a world biofuels market by virtue of cross-border investing in fuel-export platforms. In the meantime, the combination of financialization with an enabling conjuncture ('food crisis') refocuses attention on cheap land and labour in the global South, encouraging a global 'agrofuels project', depending on offshore production of alternative energy. Houtart (2010, p129) argues that while 'in the case of oil and gas, public companies have taken back their control, leaving the refining and distribution in private hands … agrofuels have entered directly into the private sector, from the production stage'. The point about financialization is that it is not simply wealthy investors like Bill Gates, James Wolfensohn, and George Soros and other financial interests (such as Louis Dreyfus, Merrill Lynch, and sovereign and pension funds) who are investing in agrofuels, but conglomerates in traditional sectors like oil, auto, chemicals and agribusiness[34] that deploy their financial resources to capitalize on the new fuel frontier.

In short, the agrofuels project combines a short-term (but unsustainable) attempt to revitalize corporate profitability with the new institutional patterns of 'offset development', sanctioned by a political-economic elite claiming to avert energy and climate crises by resorting to a new form of agro-imperialism. 'Offset development' stems from the Kyoto protocol of 'clean development', whereby Northern emissions' reduction depends on access to, and use of, Southern resources. In doing so, it reproduces a 'global ecology' whereby nature is to be managed according to a market calculus (Sachs, 1993). The consequences are the deepening of a North/South asymmetry ('ecological footprint'), and the privileging of corporate management of energy resources: converting biofuels into a fungible industrial commodity at the expense of encouraging local biofuel developments for local energy sovereignty, in accordance with the needs of food sovereignty and the reproduction of biodiversity.

## Notes

1   I am grateful to Hugh Campbell, Chris Rosin and Paul Stock for their helpful suggestions for improvement of this chapter.
2   As Merian Research (2010, p7) reports, greenwashing claims by investors of associations with environmental NGOs are often bogus. For this reason, and reasons of legitimacy under pressure from civil society, the development agencies are engaged in formulating (voluntary) codes of conduct regarding land acquisition and use.
3   In explaining financialization as a conjunctural phenomenon, Arrighi argues the recent neoliberalization of political economy is not simply a pendulum swing away from Keynesianism, but a consequence of last-ditch efforts by the US government during the 1980s to attract capital flows to the US with rising interest rates, in order to overcome the relative decline in its industrial productive capacity (Arrighi, 2007, p145). Such conditions encouraged a preference for holding capital in liquid form, and was accomplished by instituting rules promoting liberal capital markets and deregulating banking.
4   Parallel deregulation in the financial services industry thus enabled cross-over investments by banks, in addition to a process of concentration and centralization, such that between 1980 and 1998 some 8,000 bank mergers occurred, accounting for assets of over $2.4 trillion (Shattuck, 2008).
5   As of July 2008, the Standard & Poor's-Goldman Sachs Commodity Index accounted for about 63 per cent of the index fund market share, and a 32 per cent share was held by the Dow Jones-AIG index – with agricultural commodities accounting for about 30 per cent of these indices, with the rest in energy, base metals and precious metals (IATP 2008).
6   Financial speculation compounded food price inflation spiked in 2008: rice prices surging by 31 per cent on March 27, 2008, and wheat prices by 29 per cent on February 25, 2008. The New York Times (April 22, 2008) wrote: 'This price boom has attracted a torrent of new investment from Wall Street, estimated to be as much as $130 billion'; with the Commodity Futures Trading Commission noting that 'Wall Street funds control a fifth to a half of the futures contracts for commodities like corn, wheat and live cattle on Chicago, Kansas City and New York exchanges. On the Chicago exchanges … the funds make up 47 percent of long-term contracts for live hog futures, 40 percent in wheat, 36 percent in live cattle and 21 percent in corn' (quoted in Berthelot 2008).
7   Weis notes: 'Just under half of the world's total grain production (48 percent) is directly consumed by humans, while 35 percent is fed to livestock and 17 percent to biofuel

production. The surge in the latter two comes at a time when the yield gains associated with the Green Revolution have effectively maxed out, and the volume of per capita grain production on a global scale has been level since peaking in 1986' (2010, p327).

8  Thus a GRAIN researcher notes: 'Rich countries are eyeing Africa not just for a healthy return on capital, but also as an insurance policy. Food shortages and riots in 28 countries, declining water supplies, climate change and huge population growth have together made land attractive' (Vidal, 2010).

9  Weis notes that for agro-industrial producers and grain-oilseed traders 'the surging demand for fuel and feed is a strong counter-force to rising production costs, and both pressures point towards higher prices for basic foods' (2010, p328).

10  For details see ETC (2007), GRAIN (2007), and McMichael (2009b).

11  For example, in Colombia between 2001–2005, 263,000 peasant families were expropriated from 2.6 million hectares by agribusiness and/or paramilitaries interested primarily in oil palm development (Houtart, 2010, p107). Houtart claims that 60 million people risk expulsion by biofuels (Houtart, 2010, p119).

12  The minimum wage in Ethiopia is about 8 birr (39p) a day (Rice 2010).

13  FOE, Biofuels Subsidies. Available at: www.foe.org/biofuelsubsidies

14  Land-grabbing is carried out by both foreign and domestic governments, in association with either domestic or foreign investors. It has become a global phenomenon, for reasons already stated.

15  See GRAIN (2008). Roughly 20 per cent of the global land-grab is scheduled for agro-fuel crops, which, alongside of projected export food crops, constitute a new investment frontier for food, financial, energy and auto companies (Vidal, 2009, p12).

16  See also Wald et al. on Brazil, Argentina and Paraguay, in this volume.

17  For critical reviews, see the Journal of Agrarian Change, vol 39, no 6, 2008.

18  This is also the case elsewhere, such as in Southeast Asia – see, e.g. Cotula et al. (2008).

19  For example, Royal Dutch Shell is exploring a joint venture with Brazil's powerful bioethanol producer, Cosan. For Shell, this move signals growth potential to investors, and for Cosan this alliance would double ethanol production, and for Brazil it would consolidate its role as 'the world's alternative energy superpower with the potential to ship huge quantities of fuel to the US and Europe assuming a US reduction in biofuel import tariffs' (Mathiason, 2010, p43). For other details, see GRAIN (2008) and McMichael (2009b).

20  IFC expenditures in sub-Saharan Africa rose from $167 million in 2003 to $1.8 billion in 2009 (Daniel 2010, p12).

21  The World Bank's subset of investment projects notes that food crops account for 37 per cent, while fuel crops, and industrial and cash crops, each account for 21 per cent (2010, p35). The Bank's report claims that more than 70 per cent of land deals are in Africa.

22  Cf. De Schutter (2010), and Borras and Franco (2010).

23  A project manager in Mali, for example, argues 'that no country has developed economically with a large percentage of its population on farms. Small farmers with titles will either succeed or have to sell the land to finance another life … ' (MacFarquhar, 2010, pA4).

24  This is perhaps captured in the unproblematic statement of Rodney Cooke, director at the UN's International Fund for Agricultural Development (IFAD): 'I would avoid the blanket term "land-grabbing". Done the right way, these deals can bring benefits for all parties and be a tool for development' (quoted in Vidal, 2010).

25  For example, Rostow's stages of growth scenario claimed 'the revolutionary changes in agricultural productivity are an essential condition for successful take-off', and Jeffrey Sachs concretizes this trajectory by defining 'rising agricultural productivity'

as 'food production per farmer' (quoted in Weis, 2010, p315). As Weis comments: 'Enhanced productivity per worker, plant and animal are then linked in a normative way, as though inevitable aspects of development, to such things as the relative decline of the agricultural workforce, progressively cheaper food and its declining share within household expenditure, and increasing consumption of 'high-quality' foods, especially animal-derived protein' (Weis, 2010, p315–6).

26  Thus the opening paragraph of the World Bank's *World Development Report 2008* (2007, p1), reads: 'An African woman bent under the sun, weeding sorghum in an arid field with a hoe, a child strapped on her back – a vivid image of rural poverty … But others, women and men, have pursued different options to escape poverty. Some smallholders join producer organizations and contract with exporters and supermarkets to sell the vegetables they produce under irrigation. Some work as laborers for larger farmers who meet the scale economies required to supply modern food markets. Still others, move into the rural nonfarm economy, starting small enterprises selling processed foods'.

27  Especially England and the US, the two states from whose development experience 'development theory' was derived, despite the current process of 're-peasantization' in Europe (Van der Ploeg, 2009).

28  As argued elsewhere, the WTO's export regime has accelerated de-peasantization (McMichael, 2005; Araghi, 2000), for example, contributing to the transformation of Africa into a food importer, importing 25 percent of its food, and exporting high-value crops such as green beans, coffee, flowers, and biofuels. While economic theory postulates that high-value exports can assist in financing staple food imports, the food crisis revealed the limits of this scenario (McMichael, 2009c).

29  Whether agrofuels account for up to 75 per cent of food price inflation, as the World Bank reports, or somewhat less, the point is clear: that crop-derived fuels directly or indirectly contribute to rising prices. Either way, the corporate mediation of supply and demand conditions the inflation of prices. Between 2006 and 2007, US ethanol distillery corn demand increased twice as much as the increase in global demand for corn (Holt-Giménez and Kenfield, 2008, p3). The US Renewable Fuels Standards legislation (2007) empowers 'ADM, Bunge and Cargill to diversify their monopsonistic purchases to include corn for fuel as well as corn for food' (Holt-Giménez and Kenfield, 2008, p2). US corn diverted to fuel feedstock puts pressure on world grain markets, since the US produces 40 per cent of global corn supplies.

30  Thus, while Archer Daniels Midland (ADM) profits from the diversion of corn to ethanol (away from food, and from exporting to Mexico), it also profits from rising tortilla prices, since it owns a 27 per cent share of Gruma, Mexico's largest tortilla manufacturer. Further, 'ADM also owns a 40 per cent share in a joint venture with Gruma to mill and refine wheat – meaning that when Mexican consumers are forced by high tortilla prices to switch to white bread, Gruma and ADM still win' (Philpott, 2007). In fact, the 2007 'tortilla crisis' was a consequence of the diversion of white corn to cattle feed to make up in turn for the diversion of yellow feed-corn to agrofuels – with tortilla consumers forced to pay more to sustain meat consumption elsewhere (Holt-Giménez and Kenfield, 2008, p4).

31  The conversion of rainforests, peatlands, savannas, or grasslands to produce agrofuels in Brazil, Southeast Asia and the US 'creates a "biofuel carbon debt" by releasing 17 to 420 times more $CO_2$ than the annual greenhouse gas (GHG) reductions these biofuels provide by displacing fossil fuels' (Fargione, *et al.*, 2008).

32  Projections of global energy needs suggest that the ecological and social catastrophe of agrofuels is irredeemably unable to address the energy crisis of peak oil. While the UN estimates agrofuels supplying 25 per cent of energy needs over the next 15-20 years, ExxonMobil 'projects that biofuels, together with wind and solar, will contribute about

2 per cent of the world's total energy supply in 2030' (Padilla, 2007, p3). Whether ExxonMobil underestimates or not, the International Energy Agency estimates that by 2030 agrofuels will 'barely offset the yearly increase in global oil demand' (Holt-Giménez, 2007a), and all renewables, including agrofuels, will amount to only 9 per cent of global energy consumption (GRAIN, 2007, p6).

33  One estimate is that producing one litre of corn ethanol expends 1200-3600 litres of water (Houtart, 2010, p115). Deforestation also tends to reduce rainfall, and 87 per cent of deforestation in Malaysia is attributed to oil palm production (Houtart, 2010, p118): 'Agriculture is officially the most thirsty industry on the planet, consuming a staggering 72 per cent of all global freshwater at a time when the UN says 80 per cent of our water supplies are being overexploited' (Hamer and Anslow, 2010).

34  E.g. Total, Shell, BP, ExxonMobil, Petrobras, ADM, Cargill, Bunge, Monsanto, Syngenta, Dow Chemicals, Bayer, DuPont, BASF, etc. (Houtart, 2010, p131–2).

# References

Akram-Lodhi, H. (2008) '(Re)imagining agrarian relations?' *The World Development Report 2008: Agriculture for Development, Development and Change*, vol 39, no 6, pp1145–1161

Amin, S. (2003) 'World poverty: pauperization and capital accumulation', *Monthly Review*, vol 55, no 5, pp1–9

Araghi, F. (2000) 'The great global enclosure of our times: peasants and the agrarian question at the end of the twentieth century', in F. Magdoff, J. B. Foster, and F. H. Buttel (eds), *Hungry for Profit: The Agribusiness Threat to Farmers, Food, and the Environment*, Monthly Review Press, New York

Arrighi, G. (1994) *The Long Twentieth Century. Money, Power and the Origins of Our Times*, Verso, London

Arrighi, G. (2007) *Adam Smith in Beijing. Lineages of the Twenty-First Century*, Verso, London

Berthelot, J. (2008) 'Sorting the truth out from the lies about the explosion of world agricultural prices', *Solidarité*, 18 May, http://solidarite.asso.fr

Berthelot, J. (2009) 'Agribusiness' headlong flight to agrofuels and their impact on food security', *Solidarité*, http://solidarite.asso.fr

Borras, S. Jr. and Franco, J. (2010) 'From threat to opportunity? Problems with the Idea of a "code of conduct" for land-grabbing', *Yale Human Rights & Development L.J.*, vol 13, pp507–523

Burch, D., and Lawrence, G. (2009) 'Towards a third food regime: behind the transformation', *Agriculture and Human Values*, vol 26, no 4, pp267–279

Business Green (2010) 'Fossil fuel subsidies are 10 times those of renewables, figures show', *The Guardian*, August 3

Cotula, L., Dyer, N. and Vermeulen, S. (2008) *Fuelling exclusion? The Biofuels Boom and Poor People's Access to Land*, International Institute for Environment and Development (IIED), London and Food and Agricultural Organization (FAO), Rome

Daniel, S. (2009) *The Great Land Grab. Rush for World's Farmland Threatens Food Security for the Poor*, The Oakland Institute, Oakland, CA

Daniel, S. (2010) *(Mis)Investment in Agriculture. The Role of the International Finance Corporation in Global Land Grabs*, The Oakland Institute, Oakland, CA

De Schutter, O. (2010) 'Responsibly destroying the world's peasantry', 4 June, farmland-grab.org/13528

Escobar, A. (2008) *Territories of Difference. Place, Movements, Life*, redes. Duke University Press, Durham and London

ETC (2007) 'Peak soil + peak oil = peak spoils', Etc Group *Communique* #96. November/ December, pp1–14

Fargione, J., Hill J., Tilman, D., Polasky, S. and Hawthorne, P. (2008) 'Land clearing and the biofuel carbon debt', *Science*, February 7

Fraser, A. (2009) 'Harnessing agriculture for development', *Oxfam International Research Report,* September, 2009, www.oxfam.org/en/policy/harnessing-agriculture-development

Friedmann, H. and McMichael, P. (1989) 'Agriculture and the state system: the rise and fall of national agricultures, 1870 to the present', *Sociologia Ruralis*, vol 29, no 2, pp93–117

Gallagher, E. (2008) *The Gallagher Review of the Indirect Effects of Biofuels Production.* UK Government: Renewable Fuels Agency

Gillam, C. (2009) 'Investors eye global ag sector for boost', Reuters, 21 June, http://globlalandgrab.org/5695

Gouverneur, C. (2009) 'The palm oil land grab', *Le Monde Diplomatique*, 5 December, vol 5

GRAIN (2007) 'Agrofuels' (special issue). *Seedling*, July. www.grain.org/seedling/?type =68&l=0, accessed 19 February 2009

GRAIN (2008) 'Seized: The 2008 Land Grab for Food and Financial Security', *GRAIN Briefings*, October, www.grain.org/briefings/?id=212

GRAIN (2010a) 'Turning African Farmland Over to Big Business: The US's Millennium Challenge Corporation', *Seedling*, 3–5 April

GRAIN (2010b) 'The World Bank in the hot seat', *Against the Grain*, 4 May, www.grain. org/articles/?id=64, accessed August 9, 2010

Greenpeace (2007) 'How the palm oil industry is cooking the climate', www.greenpeace. org, accessed 8 January 2010

Hamer, E. and Anslow, M. (2010) 'Ten reasons why organic can feed the world', *The Ecologist*, 1 March, www.theecologist.org

Holt-Giménez, E. (2007a) 'Biofuels: myths of the agro-fuels transition', *Food First Backgrounder*, vol 13, no 2, www.foodfirst.org/node/1711, accessed 19 February 2009

Holt-Giménez, E. (2007b) 'Exploding the biofuel myths', *Le Monde Diplomatique*, July 2007, pp10–11

Holt-Giménez, E. and Kenfield, I. (2008) When 'renewable isn't sustainable': agrofuels and the inconvenient truths behind the 2007 US Energy Independence and Security Act, *Policy Brief 13,* Food First, Oakland, CA

Houtart, F. (2010) *Agrofuels. Big Profits, Ruined Lives and Ecological Destruction.* Pluto Press, London & New York

IATP (2008) *Commodities Market Speculation: The Risk to Food Security and Agriculture*, Institute for Agriculture and Trade Policy (IATP) Trade and Governance Program, 7, USA

Kaufman, F. (2010) 'The food bubble. How Wall Street starved millions and got away with it', *Harper's Magazine*, July, pp27–34

MacFarquhar, N. (2010) 'African farmers displaced as investors move in', *The New York Times*, 22 December, A1, A4

McMichael, P. (2005) 'Global development and the corporate food regime', in F. H. Buttel and P. McMichael (eds) *New Directions in the Sociology of Global Development*, Elsevier, Amsterdam

McMichael, P. (2009a) 'A food regime analysis of the world food crisis', *Agriculture and Human Values*, vol 4, pp281–295.

McMichael, P. (2009b) 'The agrofuels project at large', *Critical Sociology*, vol 35, no 6, pp825–39

McMichael, P. (2009c) 'Global food crisis: causes and prospects for policy alternatives', prepared for the UNRISD Flagship Report: *Combating Poverty and Inequality (2006–2010)*

McMichael, P. (2010) 'Agrofuels in the food regime', *The Journal of Peasant Studies*, vol 37, no 4, pp609–630

Mathiason, N. (2010) 'Shell hooks up in Brazil on biofuels', *The Guardian Weekly*, 12, February, p43

Merian Research (2010) *The Vultures of Land Grabbing. The Involvement of European Financial Companies in Large-scale Acquisition Abroad*, www.regulatefinance-fordevelopment.org

Murphy, S. (2008) 'Will free trade solve the food crisis?' *Food Ethics Council*, vol 3, no 2, www.foodethicscouncil.org

Oxfam (2007) 'Biofuelling poverty. Why the ET renewable-fuel target may be disastrous for poor people', *Oxfam Briefing Note*, November 1

Padilla, A. (2007) 'Biofuels: A New Wave of Imperialist Plunder of Third World Resources', *Special Release,* November, Pesticide Action Network Asia and the Pacific (Pan AP) and People's Coalition on Food Sovereignty (PCFS), Penang

Patel, R. (2007) *Stuffed and Starved: Markets, Power, and the Hidden Battle for the World Food System*, Portobello Books, London

Philpott, T. (2007) 'Bad wrap', *Grist*, February 22

Rama, R. (ed) (2005) *Multinational Agribusinesses*, Haworth Press, New York

Reardon, T., and Timmer, C. P. (2005) 'Transformation of Markets for Agricultural Output in Developing Countries Since 1950: How Has Thinking Changed?' in R. E. Evenson, P. Pingali and T. P. Schultz (eds), *Handbook of Agricultural Economics. Agricultural Development: Farmers, Farm Production and Farm Markets*, Elsevier Press, Oxford

Rice, A. (2009) 'Is there such a thing as agro-imperialism?' *The New York Times*, 22 November

Rice, X. (2010) 'Ethiopia – country of the silver sickle – offers land dirt cheap to farming giants', *The Guardian*, 15 January

Rosset, P. (2008) 'Food Sovereignty and the Contemporary Food Crisis', *Development*, vol 51, no 4, pp460–463

Sachs, W. (ed) (1993) *Global Ecology*, Zed Press, London

Sharma, D. (2004) 'Suicides on the farm: green revolution is turning red', *Motion Magazine*, 7 August, www.inmotionmagazine.com/opin/devsh_suicide.html

Shattuck, A. (2008) 'The financial crisis and the food crisis: two sides of the same coin', *FoodFirst*, 24 September

TNI [Transnational Institute] (2007) 'Agrofuels: Towards a reality check in nine key areas', June, from http://archive.corporateeurope.org/docs/AgrofuelsRealityCheck.pdf, accessed 19 February 2009

Van der Ploeg, J. D. (2009) *The New Peasantries: Struggles for Autonomy and Sustainability in an Era of Empire and Globalization*, Earthscan, London

Van der Ploeg, J. D. (2010) 'The food crisis, industrialized farming and the imperial regime', *Journal of Agrarian Change*, vol 10, no 1, pp98–106

Vidal, J. (2007) 'Climate change and shortages of fuel signal global food crisis', *The Guardian Weekly,* 11 September, vol 3

Vidal, J. (2009) 'Food land grab "puts world's poor at risk"', *Guardian Weekly*, 7 October

Vidal, J. (2010) 'How food and water are driving a 21st-century African land grab', *The Observer*, 7 March

Weis, T. (2007) *The Global Food Economy. The Battle for the Future of Farming.* Zed Press, London & New York

Weis, T. (2010) 'The accelerating biophysical contradictions of industrial capitalist agriculture', *Journal of Agrarian Change*, vol 10, no 3, pp315–341

Wilkinson, J. (2009) 'The emerging global biofuels market' *REVIEW*, XXXII (1), 91–112

Wittman, H., Desmarais, A., and Wiebe, N. (eds) (2010) *Food Sovereignty: International Perspectives on Theory and Practice*, Fernwood Press, Toronto

World Bank (2007) *World Development Report 2008: Agriculture for Development*, World Bank, Washington, DC

World Bank (2010) *Rising Global Interest in Farmland. Can it Yield Sustainable and Equitable Benefits?* World Bank, Washingtion DC

Xcroc (2009) 'AGRA & Monsanto & Gates, Green Washing & Poor Washing', *Crossed Crocodiles*, April 6, 2009, http://crossedcrocodiles.wordpress.com/2009/04/06/agra-monsanto-gates-green-washing-poor-washing/

Zwerdling, D. (2009) 'India's Farming 'Revolution' heading for collapse', *National Public Radio*, 13 April, www.npr.org/templates/story/story.php?storyId=102893816&ps=rs, accessed 13 August, 2010

# 6

# THE RIGHT TO FOOD: A RIGHT FOR EVERYONE

*Claire Mahon*[1]

From a right to food perspective, the past few years have been some of the bleakest in human history. According to the Food and Agriculture Organization (FAO, 2008, p6), as a result of the recent global food crisis, the number of chronically hungry people rose by 75 million in 2007, to reach a total of 923 million under-nourished people in 2008. More recently, the FAO (2009) released the projection for 2009: 1 billion and 20 million people are suffering from hunger.[2] 1 billion and 20 million. It is a number that needs to be repeated merely to grasp its magnitude. When we realize that this number represents hungry people, its gravity becomes incomprehensible.

At the core of this peak in undernourishment is the global financial crisis, which follows and partly overlaps with the other two global Fs – the global fuel crisis and the global food crisis. Because the global economic slowdown followed the 2006–08 food crisis, the impact of the food crisis impact was made even more severe. The FAO estimates that the financial crisis contributed to an 11 per cent increase in global hunger in 2009. This means that the number of hungry people is growing at a rate almost ten times that of our population growth.

Almost all the world's undernourished live in developing countries. The largest percentage of the hungry live in Asia and the Pacific, where an estimated 642 million people are suffering from chronic hunger. In sub-Saharan Africa it is 265 million, or 32 per cent, which makes for the largest prevalence of undernourish-ment relative to population size; in Latin America and the Caribbean 53 million; while the Near East and North Africa has registered the largest percentage increase in undernourishment, putting the absolute figure at 42 million. In developed coun-tries 15 million are chronically hungry.

The fact that this obscene level of hunger and malnourishment occurs in a world renowned for over-consumption is perhaps the most concerning aspect of the situ-ation. Globally, we are capable of producing far more food than is needed to feed the 6.7 billion people who currently inhabit the planet. Yet every few seconds a child dies of hunger. This is not a new phenomenon, but the situation is the worst it has ever been, and continuing to worsen.

The countries which are the most affected by the food crisis are those which are dependent upon imports for more than 40 per cent of their basic food needs. The FAO has identified 36 such countries particularly vulnerable to food insecurity, and which are most affected by the food crisis: Afghanistan, Bangladesh, Bolivia, Burundi, the Central African Republic, Chad, China, Democratic People's Republic of Korea, Democratic Republic of Congo, Ivory Coast, Ecuador, Eritrea, Ethiopia, Ghana, Guinea, Guinea-Bissau, Haiti, Iraq, Kenya, Lesotho, Liberia, Mauritania, Moldova, Nepal, Nicaragua, Republic of Congo, Sierra Leone, Somalia, Sri Lanka, Sudan, Swaziland, Tajikistan, Timor-Leste, Uganda, Vietnam and Zimbabwe. In these countries, for the vast majority of people, food accounts for at least 60 to 80 per cent of consumer spending, compared to 10 to 20 per cent in industrialized countries. A 40 per cent increase in food prices means that families in these countries must now spend their entire budget on food.

One early conclusion can be drawn: the food crisis (and the global financial and fuel crises) have exacerbated the long-lasting food insecurity plaguing the poor and most vulnerable of this world. So where is the solution? We must move beyond merely drawing attention to these dismaying figures, although this is important. We must move towards ensuring policy-makers look at the food crisis through a human-rights lens.

## What is the 'right to food'?

An understanding about what the right to food means can be found in international human rights law. It is mentioned in article 25(1) of the Universal Declaration of Human Rights and in Article 11 of the International Covenant on Economic, Social and Cultural Rights (ICESCR), which outlines the right to an adequate standard of living, including adequate food, and the fundamental right of everyone to be free from hunger. The UN Committee on Economic, Social and Cultural Rights – the body charged with monitoring the implementation of the ICESCR – explains in its General Comment No. 12 (1999) that 'the right to adequate food is realized when every man, woman and child, alone or in community with others, has physical and economic access at all times to adequate food or means for its procurement'. The first UN Special Rapporteur on the Right to Food, Professor Jean Ziegler (2006), defined the right to food as:

> the right to have regular, permanent and unrestricted access, either directly or by means of financial purchases, to quantitatively and qualitatively adequate and sufficient food corresponding to the cultural traditions of the people to which the consumer belongs, and which ensures a physical and mental, individual and collective, fulfilling and dignified life free of fear.

In the human rights jargon, when identifying the core content of the right to food we speak about availability and accessibility. Food needs to be *available* in 'a quantity and quality sufficient to satisfy the dietary needs of individuals, free from adverse substances, and *acceptable* within a given culture'; and it needs to be *accessible* 'in ways that are sustainable' (Committee on Economic, Social and Cultural Rights, 1999, para 8). As the UN Special Rapporteur concludes 'the right to food is, above all, the right to be able to feed oneself in dignity' (Ziegler, 2008, para 18). This encompasses the right to have access to resources and to the means to ensure and produce one's own subsistence, including land, small-scale irrigation and seeds, credit, technology and local and regional markets, a sufficient income and access to social security and social assistance. Access to safe drinking water is also an element of the right to food.

Most governments have ratified the international treaties which proclaim the right to food, including not only the International Covenant on Economic, Social and Cultural Rights, but also the Convention on the Rights of the Child. Consequently, these governments have accepted their obligations to *respect, protect* and *fulfil* the right to food. The UN Special Rapporteur has explained what these obligations entail in many of his reports (for example, Ziegler, 2006, paras 22–4; Ziegler *et al.*, 2011).

The *obligation to respect* is often called a negative obligation. In other words, governments should abstain from doing something. In the context of the right to food, it means governments should not arbitrarily take away people's right to food. For example, a government should not arbitrarily evict someone from his or her land especially if that land represents the main source of food production.

The *obligation to protect* has increasingly gained significance in today's globalized world. It means that governments have to enact and enforce laws aimed at preventing third parties – be they individuals, organizations or corporations – from violating the right to food. A series of related 'due process' issues arise here also, such as the duty to investigate, access to justice and the obligation to provide effective remedies. For example, pollution of a community's water supply by a company ought to be followed by government action, such as investigation, prosecution, and so on. Regulations regarding water quality, environmental protection, and other such legislation must be in place.

The *obligation to fulfil* entails two important aspects. Firstly, *facilitation*: governments must take positive action to identify vulnerable groups and to implement policies to ensure their access to adequate food by facilitating their ability to feed themselves. Introducing agrarian reform policies with the intention of improving the employment prospects of landless peasants could be such a facilitation policy. The second aspect refers to the obligation *to provide* direct assistance in situations in which people's food security is threatened for reasons beyond their control. To clarify, if a government decides to let people starve when these people have no alternative for helping themselves, this is a violation of the right to food. While the government itself might be unable to provide food for the population plagued by starvation, it nonetheless has the obligation to appeal for international humanitarian aid. Not making the appeal would be a violation of the right to food.

Understanding this three-layered obligation framework is essential. By better understanding the normative content of the right to food we can appreciate that attempts to portray the right to food as akin to requiring governments to provide everyone with caviar and champagne are simply fallacious. It also serves to show that the right to food is ultimately about the empowerment of individuals and communities, and by no means about dependency on state aid or charity.

To complete the normative framework, we should address three other important elements: progressive realization, non-discrimination and extra-territorial obligations. As the International Covenant on Economic, Social and Cultural Rights stipulates (Committee, 1999), the fulfilment of the right to food must be achieved 'progressively' and by using the 'maximum available resources'. This means that retrogressive steps are, in general, not acceptable. The current economic crisis should not be used as an excuse to cut back or dismantle social safety nets which provide those in desperate situations with minimal security, in particular access to food. In addition, human rights' law also establishes a general prohibition of discrimination, establishing there must be no delay in ensuring access to food and resources on a non-discriminatory basis.

This brings us to one of the other features of a human rights approach, taking into account how the global food crisis can affect particular groups in society. The reality is that it is women who bear the brunt of the burden of the global food crisis and the global economic crisis – it is women who struggle on a daily basis to feed their families with reduced income; women who suffer the most from disease and the effects of poor nutrition; women who are denied ownership of land, and who, in many countries around the world, forfeit any inheritance rights by virtue of their gender. Taking a human rights approach to tackling hunger, therefore, means recognizing this imbalance and addressing it.

Lastly, in terms of international human rights law obligations, there are the 'extra-territorial obligations' of governments. While the primary responsibility to ensure human rights rests with national governments, other governments cannot be seen as merely uninvolved bystanders in the international arena. Food dumping, agricultural subsidies and land-grabbing *do* affect other countries. This is the reality of our interdependent world. Through the International Covenant on Economic, Social and Cultural Rights, governments have undertaken to cooperate – without any territorial or jurisdictional limitations – to ensure the realization of the right to food. The Committee on Economic, Social and Cultural Rights (1999, para 36) states that 'state parties should take steps to respect the enjoyment of the right to food in other countries, to protect that right, to facilitate access to food and to provide the necessary aid when required'.

With this brief overview of the content of the right to food and the relevant obligations of states in mind, we move now to address the global food crisis from a human rights perspective.

## The global food crisis: a human rights perspective

Put simply, the most recent food crisis is the consequence of increased global food prices. In June 2008, the FAO food price index, based on international prices for meat, dairy, cereals, sugar and oils and fats, was 214, more than 200 per cent higher than in 2003 (FAO Food Price Index). The price of cereals, including wheat, corn and rice, which are staple foods for many people – particularly in the most food insecure parts of the world – has shown an immense growth from 98 in 2003, to 274 in July 2008. The price of cereals was 80 per cent more in 2008 than in 2003. Even though food prices decreased slightly on international commodity markets towards the end of 2008 and in 2009, on national markets, particularly in sub-Saharan Africa and South Asia, the situation remains critical (BBC News, 2009; World Bank, n.a.). More recently, the FAO (2010) has reported that the price of sugar has surpassed 30-year highs and that, in 2010 and 2011, prices will increase even more.

In a simplified form, the food crisis is the result of the imbalance between the demand for and supply of food. It is beyond the scope of this chapter to exhaustively identify all the factors that have caused food prices to skyrocket. Nonetheless, we can pinpoint some changed and changing realities which have contributed to the food crisis and which have contributed more generally to the undernourishment challenge.

Some commentators aver the 2007–08 droughts in Australia and poor crops in Vietnam, the EU and Ukraine, to be causes of the food crisis. Certainly, climate change and desertification is having its effect around the world. The World Bank (2008) and the UN Special Rapporteur on the Right to Food (de Schutter, 2008), however, considered that, during the 2008 peak of the global food crisis, good crops in other countries and increased exports offset these bad harvests, and hence this conjectural factor is unlikely to have had, on its own, a significant impact on prices.

Instead, we can point to three broad and interdependent *structural* factors:

- changing population characteristics
- the challenge of energy
- the decline in productivity growth of agriculture.

### Changing population characteristics

The globe's demographics have changed considerably in recent years. In 2007, the world population grew by 1.2 per cent: it is expected to reach 7 billion in 2012, and 9.3 billion in 2050 (Dorélien, 2008). Demand for food is estimated to double by 2030 with 20 per cent of this increase attributable to population growth. The fastest population growth is occurring in the countries most plagued by food insecurity (United Nations Population Fund, 2008).

Most reports focus on the direct impact that population growth has on the demand for food. However, demographic expansion can also have consequences on the supply side and access to food. In many developing countries, land fragmentation is a direct result of rapid population growth. Land fragmentation contributes to inefficient and destructive farming practices, which often reduce food production and result in land degradation (Sadik, 1991;

Dorélien, 2008). Meanwhile, land is increasingly being sold to wealthier land-owners, which in turn makes land distribution uneven and creates a large pool of landless labourers (Sadik, 1991). As with peasants who cannot survive from the crops planted on small patches of land, impoverished landless labourers migrate to urban areas. Research in East Africa has shown that rural households are becoming more and more dependent on wage labour, not only as a coping strategy during hunger seasons, but also as a regular strategy to meet their food needs (Dorélien, 2008).

While displacement and migration to urban areas are more often the case, cross-border movements prompted by hunger can also be observed. There is a case to be made in terms of protecting those fleeing from famine as 'refugees from hunger'. Governments continue to treat these people as illegal economic migrants, as if they have a choice in deciding to leave their countries, although there are few moral or logical reasons not to extend refugee status to those who are *forced* to flee when faced with the prospect of starvation for them and their families. As the Office of the High Commissioner for Human Rights points out, 'there is little to distinguish between a person facing death through starvation and another threatened with arbitrary execution because of her political beliefs' (Ziegler, 2007, paras 47–63).

A demographic change of a qualitative nature has also put supplementary pressure on the demand side of food. The expansion of the middle class, coupled with urbanization in developing economies (in particular China and India), has resulted in changing dietary needs: a rapid growth in demand for meat, dairy, and other high-value food products has followed. This changes the amount of food producing resources that are consumed: on average, 5kg of cereals produce 1kg of meat.

### The challenge of energy

The second factor co-responsible for the recent food crisis is the challenge of energy. The growing demand for oil due to the needs of the developed world and, increasingly, developing economies such as China and India together with speculation, and the decline of the US dollar, have pushed the oil price to new records in the recent past (Farm Foundation, 2008). Consequently, the cost of producing food has also increased: fertilizers, pesticides, transportation, packaging and processing and all other mechanized processes that use energy have become more expensive.

Meanwhile, agro-fuels have become an attractive alternative, given the high price of oil and the substantial subsidies paid to farmers for producing agro-fuels, especially by the European Union and the United States government. As the current UN Special Rapporteur on the Right to Food (de Schutter, 2008, p8) explains: 'Food and fuel compete for scarce arable land: either the land available is increased by deforestation, as seen in Brazil or in Indonesia, or less food is produced in order to fill car tanks'. According to some estimates, about

30 per cent of US maize production went into ethanol in 2008 rather than into world food markets (von Braun, 2008). The encouragement given to the agro-fuel industry to divert production of maize from food to fuel leads to increased corn prices.

The conversion of food into agro-fuel has been described as a recipe for disaster. It is estimated that it takes about 200 kg of maize to fill one tank of a car with agro-fuel (about 50 litres), which is enough food to feed one person for one year (Nicolino, 2007). Producing agro-fuels therefore creates a battle between food and fuel, leaving the poor and hungry in developing countries at the mercy of rapidly rising prices for food, land and water. If agro-industrial methods are pursued to turn food into fuel, then there are also risks that unemployment and violations of the right to food may result, unless specific measures are put in place to ensure that agro-fuels specifically contribute to the development of small-scale peasant and family farming.

This is not the first time human rights' advocates have highlighted the problem posed by biofuels, however the current food crisis has clearly highlighted the seriousness of the situation. Governments have been prompted to readjust their policies on subsidizing and promoting investment in agro-fuels. Governmental policies cannot favour investment that is detrimental to the protection and promotion of the right to food.

The rapid increase in the prices of food crops also intensifies competition over land and other natural resources. This pits peasant farmers and indigenous communities against massive agribusiness corporations and large investors who are already buying up large swathes of land or forcing peasants off their land, often through a phenomenon known as 'land-grabbing'. Land-grabbing can be understood as being the process whereby an important area of agricultural land is acquired (through long-term lease or purchase) by a foreign company or government for the production of edible crops or agro-fuels. It is argued that land-grabbing is a new self-sufficiency policy for rich net-importers of food, who have been frightened by the export restrictions or bans and increased export taxes during the recent food crisis (Cotula *et al.*, 2009, pp52–54). However, recent land-grabbing deals show a penchant not only for staples, but also for bio-fuels projects (McMichael, this volume; The Economist, 2009).

A few years ago many hailed agro-fuels as an environmentally friendly solution for energy security; today the result is quite the opposite: the move to biofuels has reduced food security and some agro-fuels significantly contribute to climate change via greenhouse gas emissions (de Schutter, 2008, pp8–9). The danger agro-fuels pose to the environment should be taken even more seriously given the general context. Climate change – in particular the issue of lack of water – is seen as the main threat to agricultural production. Experts estimate severe losses in agricultural capacity due to climate change, for example in Africa (at 17 per cent average loss and 18 per cent median loss) and Latin America (13 per cent average and 16 per cent median loss)(de Schutter, 2008, pp8–9).

## *The decline in the productivity growth of agriculture*

According to predominant economic logic, farmers as suppliers of food should be advantaged and have an incentive to produce more – particularly when food prices are rising. Thus, it might be difficult to understand why farmers are among the ones that suffer most under these conditions. A distinction, though, needs to be made: it is in reality the smallholders from poor countries that suffer most.

Since the 1980s, prices for agricultural primary commodities have generally been low. The introduction of mechanization and improved seeds in certain developing countries have not been followed by an increase in wages, and public financial support to farmers from developed countries has made it attractive for them to produce and export their crops to the developing world for lower prices than those of the local peasants. The lack of material incentive or capacity for farmers from the global South to produce much beyond subsistence levels was low also because of the retreat of the public sector from agriculture – the dismantling of subsidy programmes and disinvestment from agriculture was often promoted by international financial institutions (de Schutter, 2008, p29). Now we are waking up to the impact that such deprioritization has had. The World Bank's Independent Evaluation Group has recently found that too little has been done to support agriculture in Africa in general. Specifically, underinvestment in irrigation projects and transport infrastructure, lack of access of farmers to credit and markets, and lack of support in coping with increasing input prices such as fertilizers, has led to a decline in agricultural production (Independent Evaluation Group of the World Bank, 2007). As productivity has fallen considerably, many sub-Saharan countries have become net food importers. Thus, it is not the rather stereotypical and oft-cited causes – such as difficult environmental conditions and conflict – that are solely responsible for the current undernourishment plaguing sub-Saharan Africa.

This situation is becoming worse. In late 2010 the FAO reported that the declining size of food crops around the world, expected to worsen in 2011, has the potential to dramatically impact on global prices, risking a repeat of the 2008 crisis. According to the FAO (2010), in 2010 world cereal stocks are anticipated to shrink by six per cent, with barley plunging 35 per cent, maize 12 per cent and wheat 10 per cent. A failure to prioritize agricultural production, or, in many countries, a deliberate shift away from agriculture, has led many to be dependent on food imports. In 2010, global food import bills were set to pass the $1 trillion mark, a level not seen since food prices peaked in 2008. Meanwhile, food import bills for the world's poorest countries were predicted to rise 11 per cent in 2010 – and by 20 per cent for low-income food-deficit countries (FAO, 2010).

Given all of these constraints, and the reality of owning small patches of land, smallholders from Africa or Asia do not have the capacity to respond to the price signals of the market and benefit from increased food prices. Instead, those that do have the ability to benefit have been further distorting the global food prices through speculation in agricultural commodities. The increase in food prices in 2007 and 2008 cannot be explained without taking due account of this speculation on food and agricultural commodities. In November and December 2007, the worldwide financial markets crashed and over 1,000 billion dollars worth of investments were lost. Consequently, most of the big speculators, for example hedge funds, shifted to investing in options and futures for agricultural raw materials and staple foods.

The increase in speculation on food commodities is massive. In the year 2000, the volume of trade in agricultural products at the various stock exchanges was approximately 10 billion dollars. It was 175 billion dollars in May 2008. During just one month in January 2008, when the transfer to these markets really started, three billion new dollars were invested at the Chicago Commodity Stock Exchange.

It is difficult to calculate the degree of impact that speculative gains have had in terms of the explosion of staple food prices. World Bank economists estimate that around 37 per cent of the price explosion is due to speculation.[3] Heiner Flassbeck, Director of the Globization and Development Strategies Division at UNCTAD, judges it to be double the World Bank estimate (UNCTAD, 2008). Whatever the degree of impact, the key point is that the industry is not regulated to take into account the impact it is having, and those at the helm are not concerned. Jaques Carles, Executive Vice President of the Movement for a World Agriculture Organization (Momagri), has claimed that 'on the agricultural markets, 95 per cent of the operators are purely financial analysts. This financialization is a true drama for humanity' (Groult, 2008).[4] The current UN Special Rapporteur on the Right to Food, Professor Olivier de Schutter, explained in September 2010 how 'a significant portion of the increases in price and volatility of essential food commodities can *only* be explained by the emergence of a speculative bubble', and he emphasizes that the significant role played by large, powerful institutional investors such as hedge funds, pension funds and investment banks, 'all of which are generally unconcerned with agricultural market fundamentals', is of grave concern (de Schutter, 2010).

## Opportunities for reframing the crisis and addressing the right to food

As pointed out at the start of this chapter, from a right to food perspective, we are living in one of the bleakest times in history. The food crisis has ravaged the globe and has hit the poor and vulnerable the hardest. The economic crisis now affects the rich as well, although it is still those least able to survive who are suffering the most. Yet, even as we are experiencing this height of global hunger, we are concurrently enjoying the great riches that this world also has to offer. Clearly then, it is time for opportunities as well, a time to look at the system and make the necessary changes to reduce such contradictions.

A human rights approach to the global food situation requires a number of considerations to be taken into account. It requires the reframing of agricultural plans and policies around principles of participation (particularly of small-scale farmers), accountability, non-discrimination, transparency, empowerment, rule of law, and above all else, a focus on human dignity.

Systemic changes are needed to realize this human rights-based approach. Such changes should involve:

- prioritizing the protection of the most vulnerable (smallholder farmers);
- ensuring policies on bio-fuels respect the right to food and are geared towards true environmental protection;
- redressing the notion of food aid and reframing it in relation to the obligation of international cooperation;
- redesigning the international trading system to be compliant with human rights (including the regulation of agricultural commodities trading).

On the issue of prioritization of smallholder farmers, it is important to understand that poor rural families represent 75 per cent of the people suffering from structural hunger (Ivanic and Martin, 2008). They also suffer from increases in world food prices. Via Campesina (2008), the World Movement of Peasants' Organization, explains that:

> we also suffer from the food crisis as most small producers also have to buy food to survive. We are not the ones benefitting from the high food prices as the price at the farmgate is much lower than the price paid by consumers. Large retailers, food traders and agri-business companies are the ones profiting from the current situation.[5]

Via Campesina is right: the current food crisis not only deprives vulnerable people of their right to food, but it also benefits huge transnational corporations that monopolize the food chain, from the production, trade, processing, to the marketing and retailing of food, narrowing choices for farmers and consumers. Just ten corporations, including Aventis, Monsanto, Pioneer and Syngenta, not only control one-third of the US$ 23 billion commercial seed market, but also 80 per cent of the US$ 28 billion global pesticide market (Ziegler *et al.*, 2011). Another ten corporations, including Cargill, control 57 per cent of the total sales of the world's leading 30 retailers. In the United States, for example, 60 per cent of terminal grain-handling facilities are owned by four companies – Cargill, Cenex Harvest States, ADM and General Mills – and 82 per cent of corn exporting is concentrated in three companies – Cargill, ADM and Zen Noh (FAO, 2003).

In order to change the situation for smallholder farmers, the Bretton Woods institutions and the World Trade Organization need to change the global paradigm for agricultural policy-making and give absolute priority to investments in subsistence agriculture and local production, including irrigation, infrastructure, seeds, pesticides and the like. Peasant farmers and subsistence agriculture have been neglected for too long. The issue of the exclusion of peasants from the development process, and the neglect of their rights, should be immediately addressed.

National governments, international organizations and bilateral development agencies, should give absolute priority to investments in subsistence agriculture and local production.

In relation to the second point identified above, on the issue of bio-fuels, it is clear that environmentally damaging agro-fuels should not be produced. Only second generation agro-fuels, which do not contribute to greenhouse gas emissions, should be promoted, and there should be a moratorium on subsidizing agro-fuels that are produced from food or food producing resources, at least during periods of economic crisis, and during global food crises. Strategies must be adopted to ensure that bio-fuel production is geared towards true environmental protection and is undertaken in such a manner that the right to food is simultaneously protected.

Thirdly, there is a need to redress the notion of food aid and reframe it in relation to the obligation of international cooperation. International cooperation is not just a moral imperative, it is a legal obligation enshrined in the International Covenant on Economic, Social and Cultural Rights. States have a legal obligation to cooperate, including through financial assistance (meaning, *inter alia*, through their economic and food aid policies). Despite this, at a time when undernourishment has reached its historic peak, global food aid has hit a 20 year low (UN News Agency, 2009). International cooperation means much more than food aid. Access to seeds and fertilizers, tailored technologies, access to rural finance and markets can all be subject to cooperation. As some developed economies continue to demonstrate their success in agriculture, it becomes more imperative to ensure that their farming technology and know-how is used to assist other countries in developing better farming practices and learning how to better prioritize agriculture within their domestic economies.

Lastly, probably one of the deepest areas of systemic change, which should be prompted by the deficiencies demonstrated by the global food crisis, is the redesign of the international trading system along human rights standards. This may sound revolutionary but it is by no means a new idea: human rights advocates have already developed and advocated guidelines for such a framework (Smaller and Murphy, 2008). The starting point is the premise that the isolation of the WTO from other spheres of law, such as human rights, is unjustified as long as trade rules clearly impact on human rights. It is argued that a shift in focus is needed: trade should discipline bad practices such as dumping, excessive speculation and unchecked market power, as opposed to its current promotion of trade liberalization. The process of reaching and implementing decisions should conform to human rights standards: hence, accountability, transparency and participation of all affected groups should be at the heart of the system. As duty bearers under human rights law, governments should monitor and undertake impact assessments of the trade rules they subscribe to and revise those that conflict with human rights.

This extends to the agricultural commodities trading sector. Speculation must be regulated. UNCTAD (2008) considers that staple food prices should not be subjected to speculation on the stock exchange, but should be fixed by international agreements between producer countries and consumer countries. The UNCTAD method of regulating these agreements through buffer stocks and Stabex could be a solution. The complementary solution is to reform, drastically, the regulations for trading in futures and options through normative decisions in order to control the worst abuses.

## Conclusion

Under Article 11(2) of the International Covenant on Economic, Social and Cultural Rights, governments have recognized the fundamental right of everyone to be free from hunger. However, the reality is that the current food crisis represents a failure to meet the obligations set out to ensure an equitable distribution of world food supplies in relation to need. The food crisis also reflects the failure of national and international policies to ensure physical and economic access to food for all. The global food crisis was not an unavoidable natural disaster: it was the result of deliberate steps to downgrade the agricultural industry in countries most at need, deprioritize global food production, and destabilize the global food market. There are lessons to be learned from this, and we now have the opportunity to reframe the international system to better redress global food security through the implementation of a human rights based approach. The adoption of a human rights based approach to food and agriculture will help to ensure that those most vulnerable in the face of the global food crisis will be better protected and that their needs will be incorporated into policy-making, thus upholding the right to food as a right for everyone.

The human rights-based approach to food and agriculture is not a complex vision of utopia, however; it simply involves prioritizing human dignity and the principles of participation, accountability, non-discrimination, transparency, empowerment, and the rule of law. But in doing so, a fundamental shift in the world order will result, challenging the hegemony of the international trading system, including the WTO and the international financial institutions. People, rather than profits, would take priority, and food production for the sake of feeding people would take precedence over the skewed prioritization that policies based on comparative advantage have brought about. Laws would be designed to protect people and regulate corporations, rather than protecting corporations and regulating people. Governments would be held accountable for the starvation of their people, and for the failures in policy-making that have led to reliance on unsustainable levels of food imports. Corporations and international financial institutions would be held accountable for their involvement in undermining the realization of the right to food, and adequate regulatory systems would be embraced, including those required to end harmful commodities trading and the diversion of food production into bio-fuels.

A human rights-based approach to food and agriculture would mean that communities would 'have a say' and be involved in designing the agricultural policies that work for their society, rather than having these policies dictated by a central government – or worse, an international lender. It would mean that women would have a real opportunity to become equal players in an industry that significantly affects them, and small-scale landowners would not be passed over in favour of engagement with large corporations or foreign land-grabbers. The relationship between the various actors would fundamentally change – greater coordination would be necessary between different governments (to ensure real international cooperation); the private sector would be controlled through regulation rather than controlling through the market; and civil society would play a role in protecting the rights of those forgotten by the system. The future of food and agriculture under a human rights-based approach would fundamentally shift through the recognition that the realization of the right to food crosses borders and cuts through political priorities in multiple arenas: it is everyone's responsibility as much as everyone's right.

## Notes

1   Senior Researcher and Joint Coordinator of the Project on Economic, Social and Cultural Rights, Geneva Academy of International Humanitarian Law and Human Rights; former advisor to the UN Special Rapporteur on the Right to Food, Mr Jean Ziegler; Adjunct Clinical Professor of Law, University of Michigan Law School. With thanks to Ms Ioana Cismas for her research assistance. This chapter is based on a presentation delivered at the 2009 New Zealand Foreign Policy School, Otago.
2   Unless stated otherwise, all references to global hunger figures are from this report.
3   Personal communication from World Bank, 14 April 2008.
4   Quote translated from French.
5   Via Campesina, Press release, 4 July 2008, available at: www.europe-solidaire.org/spip.php?article10755, last accessed 31 July 2010.

## References

BBC News (2009) 'Has the food crisis gone away?' *BBC News*, 15 May

Cotula, L., Vermeulen, S., Leonard, R. and Keeley, J. (2009) *Land Grab or Development Opportunity? Agricultural Investment and International Land Deals in Africa*, IIED/FAO/ IFAD, London/Rome

de Schutter, O. (2008) *Background Note: Analysis of the World Food Crisis by the UN Special Rapporteur on the Right to Food*, 2 May, www2.ohchr.org/english/issues/food/docs/SRRTFnotefoodcrisis.pdf, accessed 20 July 2010

de Schutter, O. (2010) 'Food Commodities Speculation and Food Price Crises. Regulation to reduce the risks of price volatility', *Briefing Note by the Special Rapporteur on the Right to Food*, September, www.srfood.org/images/stories/pdf/otherdocuments/20102309_ briefing_note_02_en.pdf, accessed 1 October 2010

Dorélien, A. (2008) *Population's Role in the Current Food Crisis: Focus on East Africa*, Population Reference Bureau, Washington, DC

FAO (2003) *Trade Reforms and Food Security*. FAO, Rome, , available at www.fao.org/worldfoodsituation/FoodPricesIndex/en/, last accessed 20 July 2010

FAO (2008) *The State of Food Insecurity in the World 2008: High Food Prices and Food Security – Threats and Opportunities*, FAO, Rome

FAO (2009) *The State of Food Insecurity in the World 2009: Economic Crises – Impacts and Lessons Learned*. FAO, Rome

FAO (2010) *Food Outlook Report 2010*, November, FAO, Rome

Farm Foundation (2008) 'Complex factors drive food price increases', Press release, www.farmfoundation.org/news/articlefiles/404-FINAL%20PRESS%20RELEASE% 207-21-08.pdf, accessed July 20 2010

Groult, Y. (2008) 'La financiarisation de l'agriculture est un drame pour l'humanité: Rencontre avec Jacques Carles, Momagri', *Planetlibre*, Éte, www.planet-libre.org, accessed 8 December 2008

Independent Evaluation Group of the World Bank (2007) *The World Bank's Assistance to Agriculture in Sub-Saharan Africa: An IEG Review*, World Bank, Washington, DC

Ivanic, M., and Martin, W. (2008) 'Implications of Higher Global Food Prices for Poverty in Low-Income Countries', *World Bank Policy Research Working Paper*, April, World Bank, Washington, DC

Nicolino, F. (2007) *La Faim, la Bagnole, le Blé et Nous: Une Dénonciation des Biocarburants*, Fayard, Paris

Sadik, N. (1991) 'Population growth and the food crisis', *Food, Nutrition, and Agriculture* 1:3–6

Smaller, C., and Murphy, S. (2008) 'Bridging the Divide: A Human Rights Vision for Global Food Trade', *Background Paper for Confronting the Global Food Challenge Conference*, November, Geneva, Switzerland, www.iatp.org/iatp/publications.cfm?refid =104458, accessed 30 September 2010

The Economist (2009) 'Outsourcing's third wave: Rich food importers are acquiring vast tracts of poor countries' farmland. Is this beneficial foreign investment or neocolonialism?' *The Economist*, 23–29 May, www.economist.com/node/13692889, accessed 20 July 2010

UN Committee on Economic, Social and Cultural Rights. (1999) *General Comment 12, Right to Adequate Food*, Twentieth Session, U.N. Document E/C.12/1999/5, United Nations, New York

UN News Agency (2009) 'UN food agency warns G8 ministers of unparalleled hunger crisis as funding falls', *UN News Agency*, 12 June, www.un.org/apps/news/story.asp? NewsID=31116&Cr=WFP&Cr1=hunger, accessed 31 July 2010

UNCTAD (2008) *Trade and Development Report 2008: Commodity Prices, Capital Flows and the Financing of Investment*, UNCTAD, Geneva

United Nations Population Fund (2008) *Statement of the UNFPA on the Global Food Crisis, Population and Development*, 3 June 2008, available at www.unfpa.org/public/ News/pid/1083

von Braun, J. (2008) 'Rising Food Prices. What Should Be Done?' *IFPRI Policy Brief*, IFPRI, Washington, DC

World Bank (2008) 'Rising food prices: policy options and World Bank response', *Background Note*, April, World Bank, Washington, DC

World Bank (n.a.) 'Understanding the Crisis', www.worldbank.org/foodprices/, accessed 20 July 2010

Ziegler, J. (2006) *Economic, Social and Cultural Rights: The Right to Food*, Report of the Special Rapporteur on the right to food, E/CN.4/2006/44, Economic and Social Council, Human Rights Commission, United Nations, Geneva

Ziegler, J. (2007) *The Right to Food*, Report of the Special Rapporteur on the right to food, A/62/289, United Nations, New York

Ziegler, J. (2008) *Promotion and Protection of All Human Rights, Civil, Political, Economic, Social and Cultural Rights, Including the Right to Development*, Report of the Special Rapporteur on the right to food, A/HRC/7/5, General Assembly, Human Rights Council, United Nations, New York

Ziegler, J., Golay, C., Mahon, C. and Way, S.-A. (2011) *The Fight for the Right to Food: Lessons Learned*, Palgrave Macmillan, London.

# 7

# PLENTIFUL FOOD?
# NUTRITIOUS FOOD?

*Colin D. Butler and Jane Dixon*

Positive health requires a knowledge of man's primary constitution and of the powers of various foods, both those natural to them and those resulting from human skill. But eating alone is not enough for health. There must also be exercise, of which the effects must likewise be known. The combination of these two things makes regimen, when proper attention is given to the season of the year, the changes of the winds, the age of the individual and the situation of his home. If there is any deficiency in food or exercise the body will fall sick.

Hippocrates, 5th Century BC

## Ecology, evolution and the determinants of nutrition

The main purpose of this chapter is to present an argument for a greater understanding of the ecological factors that influence global nutrition in the past, present and future. The chapter begins by proposing that excellent nutrition in the present era requires more than the minimum quantities of calories, protein, and the known vitamins to maintain a desirable body-mass index, physical strength and cognitive capacity. While these components may be satisfactory for reasonable health, the best nutritional foundation for excellent health requires a complement of less well understood interactions between individual genetic dispositions and behaviours and food ingredients that derive from ecosystem resources.

This chapter argues that, even for affluent populations, excellent diets are elusive for four main reasons. First, the majority of the world's population exists in a milieu in which staple foods have altered enormously from those with which our ancestors evolved, so that previously low-fat protein foods, for example, now deliver unnecessary energy. Second, the bio-cultural disposition to feasting is taking place in the context of this rapidly altered food environment. Third, the nutritional ignorance which was irrelevant to our ancestors is now a source of vulnerability for those who can afford the latest processed food diets, which are accompanied by health claims of dubious merit and often based on highly reductive nutrition science. Indeed, many of the nutritional supplements which are swallowed by affluent populations have been separated from their evolutionary context, with a corresponding diminution in their nutritional values and, even, with the potential to harm. Fourth, the steady degradation of natural resources causes food yield declines along with an erosion in the nutrient qualities in those foods which are produced, to the extent that nutrient supplementation becomes inevitable.

These four interacting trends are well encapsulated by what Karl Marx referred to as the 'metabolic rift' (Foster 1999). Essentially, Marx was describing a 'robbery system' by which public goods (e.g. soil, natural resources, natural climate) are appropriated by those with the greatest economic power to produce private goods and more capital, which in turn consolidates inequality. We support this proposition by describing the implications of the metabolic rift for producer and consumer communities. If the metabolic rift does apply, then the future of nutritious food supplies is precarious indeed, with all but the very rich becoming vulnerable to shortages in the dietary diversity that is fundamental to 'good nutrition'. The chapter concludes by noting that when regional food security frameworks gloss over the distinction between energy and nutritional security, they are ignoring the complex ecology of nutritious diets.

## Nutrition security: beyond energy and nutrients

Food security is generally considered to be met by the ingestion and absorption of two principle kinds of food elements: sufficient calorific energy and adequate amounts of the known essential nutrients achieved through dietary diversity. Most industrialized countries, especially if intervening in the market economy, could comfortably guarantee a national food supply providing adequate quantities of macro and micro-nutrients. In most industrial market economies, these two characteristics are accessible for a majority of the population. While poorer sub-populations can readily access the relatively affordable energy-dense foods, many are culturally, financially and socially excluded from sufficient dietary diversity (Hawkes, 2006), and, as a result, suffer from micro-nutrient shortfalls that are injurious to health. Due to rising food prices, the number of such people rose significantly in 2008, and surpassed a billion in 2009. Since then the global financial crisis has seen this number remain at a record level, although it is important to note that the number of people deprived of a minimum supply of micronutrients is higher, affecting approximately half of the global population (Butler, in press[a]). Children and women are especially vulnerable, with undernourished children becoming disadvantaged adults, themselves frequently trapped in impoverished communities that face systematic discrimination.

The main purpose of this chapter however, is not to focus on macronutrient or micronutrient deficiency, but to discuss and integrate some lesser appreciated aspects of nutrition with perspectives from evolution, ecology and economic history. Two lineages of evidence support our focus. The first uses an evolutionary perspective to infer that large, rapid changes to the dietary milieu in which humans and their hominid precursors evolved are unlikely to be beneficial (Boyden, 1973; Ulijaszek, 2002; Cordain et al., 2005). The second stream of evidence relies on observations that diets high in diverse forms of plants (and perhaps insects) appear to be health promoting (Ogle *et al.*, 2001; Hoddinott and Yohannes, 2002). Such a diet appears safer and more beneficial than more limited fare.

## The ecology of past and modern diets

### *Three eras of food supply*

For millions of years *Homo sapiens* and its ancestors co-evolved with their environment in order to try to maximize reproduction, lifespan and well-being. These people cared about food, and they cared about reproducing themselves and their cultures. It is likely that they had a good understanding of some plant poisons, but it is unlikely that they had any intellectual understanding of nutrition, as we call it today. Human numbers slowly increased, as did the surface area they colonized and influenced. In some coastal hunter-gathering communities, resources were so abundant that highly stratified communities developed, with some even including slavery. It is plausible that elites in such areas may have been able to accumulate a sufficiently positive energy balance to become obese. Peoples living inland, especially if away from rivers and lakes, would often experience a less diverse, harder to obtain diet.

Until fairly recently the lives of most indigenous peoples have been regarded by the dominant global cultures as deprived, nasty and brutish. Perhaps some were, such as the Tierra del Fuegans observed by Darwin or the Australian aboriginals encountered by Dampier. But there is increasing recognition that the lives of at least some indigenous peoples were rich in leisure, culture, resources and of fairly good health, not least because some such peoples – especially those who rejected or did not develop agriculture – had evolved ways, some very sophisticated, to restrict their numbers and collective ecological footprint (Butler, in press[b]).

The duration of the second and third periods is, in comparison, trivial. Agriculture commenced, perhaps with the domestication of figs, no more than 12,000 years ago. In comparison to the scores of preceding millennia in which anatomically modern humans existed, agriculture started in at least six sites within a remarkably short period, and it also spread rapidly from several of these sites. For much of this second period, nutrition declined for most people, even for most elites (Gibbons, 2009); and if some hunter-gatherers did experience abundance this certainly has not been the general case for the ten millennia of agricultural experience, in which farming populations have been scarred by periodic famine, scarcity and poorer health. This fact is typically overlooked because any such reductions of life span and well-being were compensated for at the population level by increased population size, facilitating greater specialization and technological advances.

In a brief period from the mid-1800s, colonial powers – notably England, Holland, Portugal and Spain – established a system of extensive agriculture sited in a network of their colonies to supply them with the grains, protein, sugar and oil to feed their industrial workforces. In a sense they operated as absentee landlords, little concerned by the environmental, social and health impacts of their contracts for agricultural commodities. Still, for most of this period food supplies were comparatively limited: the race between 'stork and plough' was tight. Famines were periodic and common, and population obesity was scarce. This second period concluded in the late 1960s, with a time of heightened global concern over impending famines as the global population growth crested at over 2 per cent per annum.

The third period can be called the era of agricultural intensification, marked by the green revolution and by the development of 'landless farms', inexorably breaking the nexus between locally grown stock and locally grown grass and other forms of feed (Naylor *et al.*, 2005). Landless farms facilitated extremely dense animal populations, especially of monogastric chickens and pigs and, to a lesser extent, digastric (methane-producing) cattle, goats and sheep. Supplying the feed for these animals has become a global business, with a particularly large trade of grain and soy between fields in South America and concentrated animal feeding operations in Asia (Wald *et al.*, this volume).

This third phase has greatly expanded the global supply of meat and calories and, for reasons of environmental degradation and wealthy consumer demand described below, made it harder for many population groups to access affordable grains and pulses. Not only has the relative proportion of available foods changed over the millennia, so too has the content of foods. Some of these latter changes have been rapid. For example, the fat content in intensively farmed animals whose exercise capacity is extremely restricted is much higher than the 'extensively farmed' animals from fifty years ago; and higher still than in wild animals, even at their time of peak fat storage at the start of winter (Cordain *et al.*, 2005). This is documented for chickens (see figure 7.1) (Wang *et al.*, 2010). The quality of that fat, including its quantities and ratio of essential fatty acids has also changed (Wang *et al.*, 2010), almost certainly in adverse ways.

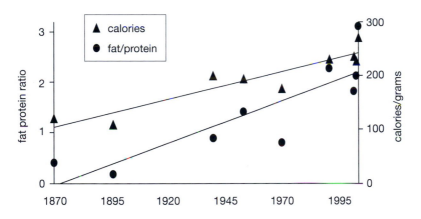

Source: adapted from Wang (2010)

*Figure 7.1* The percentage of fat found in chickens, now mostly reared in industrial conditions, has risen substantially.

Given this context, knowledge of what we can reasonably expect of food yields and of the nutritional quality of foods and their destinations is of great importance. Using WHO Expert Committee 'healthy diet and population nutrient intake goals', Hawkesworth et al., (2010) assess the performance of the global agricultural system against widely accepted nutritional goals for daily intake (see table 7.1). They cite the latest evidence supporting the health-protective advantages of this targeted food composition, as well as evidence of associations between diet-related disease and the penalty when recommended intakes are exceeded for saturated fatty acid, free sugars, and red meat.

For three decades now, we have seen that the trend of intensive food producing and processing using the cheapest possible ingredients has harmed human health (Alston et al., 2008). Some writers have suggested that a decline in nutrition coincided with the start of widespread availability of processed foods, mediated especially through high intake of sugar (Clayton and Rowbotham, 2008) The food services sector plays a critical role in shaping consumer demand, but for a narrow range of foods (Hawkes, 2008) rather than more desirable dietary diversity; it is now thought that processed foods with their 'quick energy' displace a desire to eat plant foods. Disturbingly, the annual growth in sales of processed foods is now much higher in developing countries than in rich countries (Blouin et al., 2009).

The scale of sub-optimal nutrition is large, increasing and has several forms. One may be due to an insufficient intake of flavonoids and other anti-oxidants, which are found in many diverse diets. Another form may occur through the ingestion of excess saturated fatty acids, though the effects of different forms of fats is increasingly contested (Stamler, 2010), an example of 'nutritional cacophony' (Fischler, 1993). As a result of the consumption of higher proportions of processed food, there is also wide appreciation that many people (up to a billion or more) ingest excessive sodium – a major cause of hypertension and ischaemic heart disease (Cordain et al., 2005).

Table 7.1 Nutrient intake targets for a healthy diet

| Item | Target[a] |
| --- | --- |
| Complex carbohydrates | 55–75% |
| Simple and naturally occurring sugars | <10% |
| SFAs | <10% |
| Protein | 10–15% |
| Fruit and Vegetables | >400 gm/day |

Source: Hawkesworth et al. (2010)

Note
a   Figures given as per cent refer to that item's respective proportion of dietary energy intake

## *Overeating, under-exercising and evolution*

Periodic scarcity is likely to have contributed to the evolution of behaviour in which essential nutrients such as elements and fat-soluble vitamins, as well as calories, could be ingested in excess (where available), stored, and thus made available for later physiological use. As societies evolved, some developed physical methods of food and nutrient preservation, such as the salting or drying of fish and meat. Some agricultural cultures started to bury and delay the harvest of root crops, and to store grain and other foods (e.g. yams) in specially built houses. Some coastal peoples developed techniques to store living fish in man-made or natural rock enclosures. If a nomadic people found a hive of honey, then gorging made ecological sense. In places and times of periodic or recurring scarcity, populations that consistently squandered food and other resources are unlikely to have been as successful as their more pragmatic neighbours.

Ulijaszek (2007, 2002) has also pointed out how evolution of this modern environment has outpaced human bio-cultural dispositions, and thus the common human inability to disregard or override their wish to feast is, in part, responsible for the global obesity epidemic. But there is another dimension too: in the past, occasional feasting was unlikely to be harmful, and it probably assisted survival. Today, feasting (or at least the habitual consumption of more calories than are expended) is not only common, but it often also involves the ingestion of foods which are substantially different – especially through being more calorie-dense – than foods available to our ancestors.

A possible explanation for the observation that many people eat far more calories than they require for their daily metabolism is that the ancient and often rational practice of gorging, whenever there is an opportunity is, if not 'hard-wired', at least tempting to many people. This characteristic is likely to be strongly favoured by natural selection, though influenced, modified and in some cases reinforced by new cultural norms acquired through spending time abroad as part of diasporic communities (Dixon and Jamieson, 2005), and by the cultural normalization that has been underway for 40 years through 'fast food' promotions of abundant servings at low cost: the 'supersize me' phenomenon.

In short, the evolution of these trends in a large part reflects the evolutionary forces which led so many of our ancestors to feast when they had the opportunity, and, as long as it tasted good, not to worry about its nutritional properties. These forces, which we have suggested above are in part 'hard-wired', have not only driven more or less unconscious acceptance of the global feast available in supermarkets and fast food outlets, but have also driven the supply of those foods. This also means that, in the main, advocates for agriculture have not only a commercial motivation for promoting a 'more is better' global dietary provision, but share the same unconscious belief that progress and abundance are inseparable value systems.

Of course, the reasons for modern obesity are more complex than opportunistic gorging. Another is that survival in most modern societies requires the expenditure of fewer calories than in the environment in which hominids evolved. Some of our hunter-gatherer ancestors may have had leisure-rich lives, but their daily activities still involved more activity and energy expenditure than is generally the case today. Most of our agrarian ancestors (apart from a small privileged minority) are likely to have worked harder than either modern people or hunter-gatherers, and until recently had distinctly worse nutrition. Many still do.

### Nutritional knowledge and nutrition science

Until recently, most humans have, and have had, little conscious understanding that different foods contain different quantities, qualities and ratios of nutrients. Even so, learned and instinctive preferences and aversions to certain foods modify their attractiveness for humans, other animals including invertebrates, and perhaps even microbes. Thousands of years ago, some cultures had evolved plant-based cuisines which provided complementary amino acids, such as rice and dahl, and corn and beans. Hundreds of years ago, a primitive understanding of vitamins existed, particularly among seafarers. Of course, the active dietary constituents were not then isolated, but custom ensured that basic nutrition was supplied for most, apart from at times of famine, at times of conflict, or at sea. Some populations had also learned how to safely use foods in ways to reduce their toxicity, such as the proper preparation of the cyanide-containing root crop, cassava.

The recent discovery of bones showing evidence of butchery using stone tools has pushed back the date of sophisticated hunting by hominids to more than three million years ago. These human predecessors seem to have greatly valued the meat and marrow of the large animals they killed in cooperative groups, not least because they transported their stones six kilometres from the site where they are naturally found to the place where the butchery occurred. Their preference for such foods incurred other risks and costs, including being vulnerable to attack by large carnivores (Braun, 2010).

This quest for meat and marrow did not occur because these hominids had read any food guidelines. However this practice commenced (some non-hominid primates, such as chimpanzees, also occasionally kill for meat), it is clear that hunting, meat eating and probably the status of being a successful hunter, co-evolved. Even today, the killing and consumption of bush meat has high status in some human societies. Such status is unlikely to have developed and persisted for millions of years if the trophy brought back did not also possess good nutritional and group-affirming properties, not easily substituted via other strategies. One of these properties may have been the quantity of nutritious food, as the killing of a large animal, especially in a hot climate, would justify feasting upon it by several people, as it would otherwise quickly deteriorate. Meat was also likely to be prized by hunter-gatherers because it conferred strength and stamina as much as it provided calories. Although unknown to our ancestors, meat contains the eight essential amino acids, and is also rich in zinc, vitamin B12, and in some cases may provide iodine. A particularly valued trait, especially of red meat, is its energy-facilitating iron, which occurs in a molecular form that is much easier to absorb than that in plant foods. In some environments, especially away from the coast, this combination of essential nutrients would otherwise be very difficult to obtain.

It is thus plausible that the ancient consumption of animal parts would sometimes have had a highly valued restorative effect, for example if eaten after a lengthy period of a purely vegetal diet. On the other hand, the regular consumption of animal products would maintain but not necessarily revive health. When replete, vitamin B12 stocks in humans can be slowly released over more than a year without deficiency. Thus, reasonable B12 status can be maintained for many months even in the absence of any animal products. Other nutrients, especially those supplied mainly or exclusively by plants, require much more frequent ingestion, such as vitamin C. 'Gatherer-hunters' may be a more appropriate term, but virtually all ancient humans consumed some animal products. Such products would often not include muscle meat; shellfish, fish, eggs, offal and blood also provided valuable nutrients, including iron and vitamin B12.

Knowledge of nutrition expanded in the second food era (Cannon, 2005). The nineteenth century was notable for the emergence of professional nutrition science, which arose from the discipline of chemistry. Much has now been written about how the discovery of protein, carbohydrates and vitamins influenced company product lines and marketing strategies (think Kellogg); played a role in the establishment of the Food and Agriculture Organization; informed the directions of inter-governmental negotiations on trade and assistance packages; and led relatively educated consumers to change their diets quite dramatically (Scrinis, 2008; Dixon, 2009). At times, nutrition science has prematurely recommended nutrient intakes with insufficient consideration of the possible dangers of over-consuming, including the FAO's 'protein crisis' of the 1950s, later called a 'protein fiasco' (Cannon, 2005). We are not proposing that people had any understanding that the foods which formed the staple hunter-gatherer diet were generally nutritious in comparison to today's diet. However, we suggest that this widespread nutritional ignorance has unwittingly conditioned the global population to accept, with little question, the mass-produced food to which so many people are now accustomed.

### Climate change, food yields and nutrient quality

The Food and Agriculture Organization (FAO) estimates, perhaps optimistically (Butler, 2010), that in 2050 there will be more than sufficient calorific energy in the global food supply (3050 kcal per person per day), although at the same time the agency notes that the world will by then need 70 to 100 per cent more food than today. (It is unclear if these food data are adjusted to exclude crops grown for ethanol and other forms of bio-energy, such as maize, soy, sugar and palm oil grown (Butler, 2009)). This unarguably reflects an understanding of a distinction between possibilities for dietary energy and quality. What it also reflects, however, is an expectation of a 'business-as-usual' approach, in which the recent trend of landless farms and the intensification of animal production – relying on fossil fuel-intensive land clearing and feed shipping – will continue.

Analysts working within the climate change and agriculture field already argue that dietary quality is deteriorating, without considering the additional complexities of changed food processing. They are further concerned that bio-physical resources will reach critical thresholds. Climate change is anticipated to affect all aspects of food security: food yields, availability, quality, access and utilization values (Butler, 2010; Edwards *et al.*, 2011). Most worrying is the fact that grain yields are growing at slower rates. Of all grain crops, rice yields are increasing the most slowly – which is of particular significance given rice is a staple for about one third of the world population. In the case of maize, much yield increase is being diverted to ethanol production, in an unsustainable attempt to supplement petroleum supplies and mitigate greenhouse gas emissions. As we write, the world is again experiencing record food prices, the fundamental cause of which is insufficient production to meet an inexorably increasing demand.

Soil nutrient depletion also harms nutrition. Elevated carbon dioxide levels are predicted to reduce the protein content of wheat, and perhaps of other grains. Additional concerning indicators of inferior dietary quality due to climate change are more toxins in heat- and rain-affected crops, and increased food spoilage due to poor storage in hot or wet weather.

In summary, the global agricultural system is using limited supplies of fossil fuels and other resources – which are becoming more scarce – to produce and transport nutrients around the world, in order to meet the growing demand of a population which is, at the population level, poorly equipped to recognize when it has had too much of a good thing. We acknowledge that a substantial part of this global trade in nutrients has been of benefit to health, such as by enabling a higher consumption of nutrient-dense food by populations in low income countries. Yet, even there, consumption by many is likely to increasingly exceed thresholds of dietary harm, not to mention exert a growing cost to the Earth system.

## The metabolic rift and producer and consumer community interactions

A feature of the third food era is that the numbers of calorically undernourished and calorically over-nourished are unprecedented. At the same time, never before have so many people been so dependent on others to provide food for them. As part of this paradox, many small farmers who could supply their own food are in debt, impoverished and chronically undernourished. As producer communities become more enmeshed in the global food supply to meet their own dietary needs, two troubling ancillary trends have been noted (Blouin *et al.*, 2009): global food companies have been able to make processed foods of dubious nutritional worth available and affordable for huge numbers of the world's population; while global supermarkets and fast food chains have made such foods not only highly accessible but also highly acceptable.

The interrelationships between these dynamics are accelerating a phenomenon identified by Marx in the 1880s as 'the metabolic rift'. The appropriation of agricultural lands for purposes of capital accumulation, rather than food production, amounted in his terms to 'a robbery system' exploiting labour and soils: the two factors considered by Marx as the most basic foundations of wealth (Foster, 1999). Under the agrarian system, direct, rapid and locally apparent feedbacks existed between nutrient cycles, plant and animal yields and human life and death. Marx argued that, to ensure its survival in the context of exploited soil, capitalism had three main options: (i) to expand into new nutrient-rich environments; (ii) to apply technologies to restore soil fertility, such as the use of synthetic fertilizers in existing locales; and (iii) to develop new plant and animal varieties which could survive nutrient-depleted states.

Transnational corporations have indeed followed this script according to the many current applications of 'the metabolic rift' paradigm. However, Marx intended that the concept operate more dialectically so that actions by producer-traders and consumers could be simultaneously analysed. He foresaw that the key to exploitation, as opposed to sustainable production-consumption, lay in the disengagement of relationships, whether between different social classes or between people and their environments: between society and nature, as he put it.

In these next paragraphs we turn to some of the manifestations of that disengagement. We begin from the premise that large numbers of food consumers are, for complex bio-cultural reasons, driven by their evolutionary dispositions towards particular dietary practices (such as feasting) in the food-enriched environments which many now inhabit. As these combined realities intersect with a dependence on others, often unknown producers and companies located many thousands of kilometres away, the feedbacks between actions and resources become extremely attenuated (Campbell, 2009). The absence of feedbacks is a multi-dimensional and multi-scalar process: it implicates culinary cultures, ecosystems and human communities and it ranges from the individual metabolism to the planetary system.

For example, in search of the elusive goal of optimal nutrition, large numbers of people have been induced, often through advertising, to consume large quantities of synthetic vitamins, either as pills or hidden as fortified foods. Many others follow faddist diets. This is despite the substantial evidence that the consumption of a wide range of micronutrient-rich foods, especially fruit and vegetables, provides people with compounds which promote improved health, including by lowering risks of cancer and heart disease. Yet, although vitamin deficiencies are harmful, the evidence that high vitamin doses are beneficial is far frailer than promoters would wish. In fact, the strongest epidemiological evidence concerning supplementary ingestion of vitamins and even minerals is that they harm health (Lawlor et al., 2004). Overall, the evidence for benefits of supplementation with isolated molecules is reductionist. The uncritical mass consumption of such supplements by apparently healthy people truly justifies the term "vitamania" (Apple, 1996).

Such public health concerns are consistently overlooked by giant food processors, like Nestlé and other actors in the food services sector including supermarket and fast-food chains, even though they position themselves as strongly concerned with health and wellness (Burch and Lawrence, 2009). In the main this is a defensive strategy designed to reclaim legitimacy eroded through successive food scares and the accumulating evidence that a sick food system causes sick populations (Dixon and Banwell, in press).

It also capitalizes on the fears of 'the worried well': the middle classes who are increasingly obsessed with health protection and attempts to fend off disease through diet. For them, at the moment, anti-oxidant rich diets are the holy grail. Just as sales of margarine and chicken escalated in the 1970s when nutrition scientists established a link between heart disease and high fat diets, foods associated with claimed anti-oxidant properties command high prices in the global marketplace. Interestingly anti-oxidants are among the harder to measure nutrients and, as such, are more open to nutritional cacophony and hence to exploitation in the marketplace than longer recognized nutrients.

In a recent special issue of *Globalizations*, half a dozen case studies illustrated the cascade of negative effects that can result when domestic agriculture is reoriented to produce export foods that are prized by wealthy consumers for their anti-oxidant and other nutrient values – including tea, chocolate and exotic fruits and vegetables (Cooke *et al.*, 2008). Fish is an exemplar case, following its promotion as being rich in beneficial omega-3 fatty acid. Many national dietary guidelines advocate at least two servings of, preferably oily, fish a week; thus its demand, even in traditionally meat-eating countries, is escalating. In Australia, the volumes of imported fish increase each year, driven by price signals. Thailand, whose fishing industry is underpinned by cheap and exploited labour, is a favoured source of those fish. Thai fisherpeople are now reliant for their incomes, and their access to food, on contracts with large transnational export firms, and this has been encouraged by the Thai government as part of its drive to become the 'kitchen to the world'.

A move away from food self-reliance to income generation, through the export of embedded nutrients to cosmopolitan centres, extracts a high cost to producer ecosystems and to human health and well-being, both for producers and even consumers. Gambia's recent development of rice exports, encouraged by development agencies, has had a significant negative effect on the environment and society, with Gambians now reliant on imported rice for their own consumption. Their situation compares less favourably to Mali, whose citizens continue to produce more of their own rice and who were able to switch to home-grown sorghum (Moseley *et al.*, 2010).

Because of different bioregional and social group capacities for food production, global trade in food is inevitable. But we need to be aware of the many sides to this form of food provisioning. There have been numerous other studies of the mixed fortunes in the fruit, chocolate and tea sectors – especially where Fair Trade has been anticipated to both improve environmental practices and to improve wages and the working conditions of labourers. While this can indeed be the result, it can also mean precarious working lives which can end abruptly when contracts are not renewed; immiseration for smaller producers unable to upscale sufficiently to obtain the contracts; and displacement of self-provisioning of food by marketplace reliance.

In summary, whilst the current global agro-food system complex produces ample calories, together with macro and micronutrients, to alleviate malnutrition (though not necessarily to provide optimal nutrition), the system extracts an extraordinarily high cost from the social and bio-physical environment. It also has a high ethical price, including the treatment of low-paid, often exploited agricultural workers, and the cruelty of intensive animal farming. Considering the scale of this harm, the benefit to global health of the current food system is far less than is often assumed. We agree with Hawkesworth and colleagues (2010) that a distinction should be made between 'feeding the world' and 'feeding the world healthily' (Hawkesworth *et al.* 2010). It seems that the current global food and agricultural system can do the former, though we add that this is subject to a fairer distribution of food 'entitlement' (Pritchard, this volume).

To this point, our chapter intimates points of leverage. Even if we are partly 'hard-wired' to feast when food supplies are plentiful, we are also plastic. Fads come and go: for example, the recent move away from margarine with trans-fats, and an embrace of more organic, health and sustainability arguments to reduce meat, especially grain-fed. However, while these are the marketplace and social-movement fostered 'options' for the middle-classes, a more comprehensive safety net approach is required for the bulk of the world's population. Furthermore, if we are to advance and to sustain nutritional advances for all peoples, an ecological approach is essential in which food policy, energy policy and climate change policy are not only simultaneously considered, but where an integrated policy approach acknowledges the bio-cultural and sociocultural legacy of human history.

If we value social and bio-diversity, ecological feedbacks and a fair global food system, it appears logical to reassess the contribution that multiple forms of food provisioning play in pursuing nutritional security. These include the self-provisioning practices of hunter-gatherers as exemplified by modern day gleaners and community and home gardeners; the commercial market, accessed by cash or financial credit transactions; charity and aid, often entailing donors some considerable distance away providing one-off assistance in times of calamity; and development assistance brokered between governments and multilateral agencies.

Each has its place depending on circumstances, and each has downsides. Self-provisioning in bio-spherically difficult environments does not guarantee desired dietary diversity, and hence there has to be reliance on others for food supplies. Food provided through aid packages, development assistance and global free trade is essential in many places and at particular times, but each of these 'foodways' has often supplied plentiful calories but not plentiful nutrition, perhaps because they have also been a mechanism for offloading surpluses from developed world producers.

In context, where there is no sure route to nutritional security, governments need to closely interrogate the many food security agreements that they are signing up to. In the Asia-Pacific region alone, the following frameworks are among the many more to have been promulgated in the last two years alone: the ASEAN-FAO Regional Conference on Food Security; the ASEAN Multisectoral Framework on Climate Change and Food Security; the APEC Food System; and the ASEAN Plus Three Roundtable on food security cooperation strategies. In an encouraging sign, there is increasing recognition in these documents that producer-community survival is dependent on environmental sustainability and fair terms of trade. Thus, development concerns are being aligned with environmental concerns, and, in reference back to the metabolic rift concept, feedbacks between social development and natural resource depletion are being considered. However, most adopt without question the 1996 World Food Summit definition which does not acknowledge that it is possible to have energy security and not nutritional security. It appears that what is amenable to measurement (caloric energy) 'counts' in a way that insights from ecological history do not.

## Conclusion

Life expectancy has increased significantly in middle- and high-income countries since the nineteenth century, in part because of improved nutrition, together with better housing, vaccinations and other social and medical advances. As recently as the mid 1960s, there were widespread, well-grounded fears of vast famines in low-income countries. In the main, these famines did not eventuate, but the number of chronically undernourished people is today estimated at more than a billion, and in some countries more than 40 per cent of children are chronically undernourished. On the other hand, at least a billion people live in an environment where the food they consume has excess calories, but not necessarily sufficient micronutrients.

While there has been an enormous expansion in the production and consumption of food, especially of animal products, fat and carbohydrates, the production and consumption of the nutritious food which humans evolved to eat has not kept pace. The current maldistribution and inequitable consumption of dietary diversity is likely to become more pronounced as deleterious climate change becomes further entrenched, and as energy prices rise. We propose that the malconsumption of dietary diversity is partly a result of our evolutionarily-driven propensity to feast, interacting with the global maldistribution of nutritional diversity, which in turn is partly attributed to a global corporatized food system. This maldistribution is not a result of nutritional ignorance because we note that people did not have to be nutritionally literate to be nutritiously fed. The quote from Hippocrates at the beginning of our chapter indicates a keen awareness that eating foods that require physical activity – that is, local foods that were also appropriate to the natural surroundings (seasonal, etc.) – are health promoting. Rather, malproduction, distribution and consumption are features of a metabolic rift which is the result of, and in turn fosters, disengagements between the environment, food producers and consumers.

This chapter highlights the uneven way in which nutritional and health insecurity is playing out in different parts of the world; a situation which is deteriorating as wealthy consumers seek health protective foods. We acknowledge that, historically, a substantial part of the global trade in nutrients has been of benefit to health through enabling a higher consumption of macronutrient-dense food by populations in low-income countries. Yet, even there, consumption by many is likely to increasingly exceed thresholds of dietary harm, not to mention exert a growing cost to the Earth's system. Unless we ensure that health, sustainability and justice considerations shape the food and agricultural system as growing numbers suggest, and which has pertained for millennia, then the agri-industrial complex will further erode public health, as well as many other forms of public good.

## References

Alston, J., Sumner, D. and Vosti, S. (2008) 'Farm subsidies and obesity in the United States: national evidence and international comparisons', *Food Policy,* vol 33, no 6, pp470–479

Apple, R. (ed) (1996) *Vitamania. Vitamins in American culture,* Rutgers University Press, New Brunswick, NJ

Blouin, C., Chopra, M., and Van Der Hoeven, R. (2009) 'Trade and social determinants of health', *The Lancet,* vol 373, pp502–507

Boyden, S. V. (1973) 'Evolution and health', *The Ecologist,* vol 3, pp304–309

Braun, D. R. (2010) 'Australopithecine butchers', *Nature,* 466, p828

Burch, D. and Lawrence, G. (2009) 'The "Wellness" phenomenon: implications for global agrifood systems', in G. Lawrence, K. Lyons, and T. Wallington (eds) *Food Security, Nutrition and Sustainability,* Earthscan, London

Butler, C. D. (2009) 'Food security in the Asia-Pacific: Malthus, limits and environmental challenges', *Asia Pacific Journal of Clinical Nutrition,* vol 18, pp 577–584

Butler, C. D. (2010) 'Climate change, crop yields, and the future', *SCN News,* vol 38, pp18–25

Butler, C. D. (in press[a]) 'Malnutrition and under-nutrition', in M. Juergensmeyer and H. Anheier (eds) *Encyclopedia of Global Studies,* Sage Publications, Thousand Oaks, CA

Butler, C. D. (in press[b]) 'Population Trends and the Environment', in R. H. Friis (ed) *Praeger Handbook of Environmental Health,* Praegar, Westport, CT

Campbell, H. (2009) 'Breaking new ground in food regime theory: corporate environmentalism, ecological feedbacks and the "food from somewhere" regime', *Agriculture and Human Values,* vol 26, pp309–319

Cannon, G. (2005) 'The rise and fall of dietetics and of nutrition science', 4000 BCE–2000 CE. *Public Health Nutrition,* vol 8, pp701–705

Clayton, P. and Rowbotham, J. (2008) 'An unsuitable and degraded diet? Part one: public health lessons from the mid-Victorian working class diet', *Journal of the Royal Society of Medicine,* vol 101, pp282–289

Cooke, A., Curran, S., Linton, A. and Shrank, A. (2008) 'Introduction', *Agriculture, Trade, and the Global Governance of Food,* vol 5, pp99–106

Cordain, L., Eaton, S. B., Sebastian, A., Mann, N., Lindeburg, S., Watkins, B. A., O'Keefe, J. H. and Brand-Miller, J. (2005) 'Origins and evolution of the Western diet: health implications for the 21st century', *American Journal of Clinical Nutrition,* vol 81, pp50–54

Dixon, J. (2009) 'From the imperial to the empty calorie: how nutrition relations underpin food regime transitions', *Agriculture and Human Values*, vol 26, pp321–331

Dixon, J. and Banwell, C. (in press) 'Choice editing for the environment: A corporate response to the impacts of hyper-consumption on the environment', in S. Lockie and T. Meacham (eds) *Risk and Social Theory in Environmental Management*, CSIRO Publishing

Dixon, J. and Jamieson, C. (2005) 'The cross-Pacific chicken: Tourism, migration and chicken consumption in the Cook Islands', in N. Fold and B. Pritchard (eds) *Cross-Continental Food Chains*, Routledge, London

Edwards, F., Dixon, J., Friel, S., Hall, G., Larsen, K., Lockie, S., Wood, B., Lawrence, M., Hanigan, I., Hogan, A. and Hattersleyl, L. (2011) 'Climate change adaptation at the intersection of food and health', *Asia-Pacific Journal of Public Health*, vol 23 (2), suppl 91S–104S

Fischler, C. (1993) 'A nutritional cacophony or the crisis of food selection in affluent societies', in P. Leatherwood, M. Horisberger and W. James (eds) *For a Better Nutrition in the 21st Century*, Vevey/Raven Press, New York

Foster, J. B. (1999) 'Marx's theory of the metabolic rift: classical foundations for environmental sociology', *The American Journal of Sociology*, vol 105, pp366–405

Gibbons, A. (2009) 'Civilization's cost: the decline and fall of human health', *Science*, vol 324, p588

Hawkes, C. (2006) 'Uneven dietary development: linking the policies and processes of globalization with the nutrition transition, obesity and diet-related chronic diseases', *Globalization and Health*, vol 2, (e) 4

Hawkes, C. (2008) 'Dietary implications of supermarket development: A global perspective', *Development Policy Review*, vol 26, pp657–692

Hawkesworth, S., Dangour, A. D., Johnston, D., Lock, K., Poole, N., Rushton, J., Uauy, R. and Waage, J. (2010) 'Feeding the world healthily: the challenge of measuring the effects of agriculture on health', *Philosophical Transactions of the Royal Society B*, vol 365, pp3083–3097

Hoddinott, J. and Yohannes, Y. (2002) 'Dietary diversity as a food security indicator', *FCND Discussion Paper*, International Food Policy Research Institute, Washington, DC

Lawlor, D. A., Smith, G. D., Bruckdorfer, K. R., Kundu, D. and Ebrahim, S. (2004) 'Those confounded vitamins: what can we learn from the differences between observational versus randomised trial evidence?' *The Lancet*, vol 363, pp1724–1727

Moseley, W., Carney, J. and Becker, L. (2010) 'Neoliberal policy, rural livelihoods, and urban food security in West Africa: A comparative study of The Gambia, Cote d'Ivoire, and Mali', *Proceedings of the National Academy of Science of the US*, vol 107, pp5774–5779

Naylor, R., Steinfeld, H., Falcon, W., Galloway, J., Smil, V., Bradford, E., Alder, J. and Mooney, H. (2005) 'Losing the links between livestock and land', *Science*, vol 310, pp1621–1622

Ogle, B., Hung, P. and Tuyet, H. (2001) 'Significance of wild vegetables in micronutrient intakes of women in Vietnam: an analysis of food variety', *Asia Pacific Journal of Clinical Nutrition*, vol 10, pp21–30

Scrinis, G. (2008) 'On the Ideology of Nutritionism', *Gastronomica: The Journal of Food and Culture*, vol 8, pp39–48

Stamler, J. (2010) 'Diet-heart: a problematic revisit', *American Journal of Clinical Nutrition*, vol 91, pp502–509

Ulijaszek, S. (2002) 'Human eating behaviour in an evolutionary ecological context', *Proceedings of the Nutrition Society,* vol 61, pp517–526

Ulijaszek, S. (2007) 'Obesity: a disorder of convenience', *Obesity Reviews,* vol 8, pp183–187

Wang, Y., Lehane, C., Ghebremeskel, K. and Crawford, M. (2010) 'Modern organic and broiler chickens sold for human consumption provide more calories from fat than protein', *Public Health Nutrition,* vol 13, pp400–408.

# 8

# A UTOPIAN PERSPECTIVE ON GLOBAL FOOD SECURITY

*Paul Stock and Michael Carolan*

One of us (Michael) was recently involved in bringing famed environmentalist Lester Brown (founder of the Worldwatch Institute and more recently the Earth Policy Institute) onto campus at Colorado State University to give a series of public lectures. When word began to spread through the agriculture community about Lester's invitation, the pushback was palpable and immediate. Colorado State University, you see, is a Land Grant University, with a particularly strong history in livestock production. Colorado is home to some of the largest beef CAFOs (confined animal feeding operations) in the world (the largest in the state has a one-time feeding capacity of 811,000 head of cattle). So the University has a close working relationship with large-scale beef feeders. The outrage stemmed from the belief that the University, and specifically the College of Agriculture (where the invitation originated), had invited someone who preaches eating lower in the food chain; someone, in other words, who was unabashedly anti-feedlot. Chests were puffed as not-too-veiled threats were made about what would happen if his invitation was not retracted. Ultimately, Lester made his trip to Fort Collins, gave his talks (he crafted his message carefully in recognition of his audience), and left.

The criticisms levelled at Lester and his messages are common, we've both heard them many times before. To boil it down to its core, the argument often goes something like this: the problem of global food security needs science-based solutions, not pie-in-the-sky theories – 'people live off fish not fantasies' to adapt the famous proverb. That's the critique, at its heart, that conventional agriculture directs at the competing alternatives (organic, local, etc.): that they are too long in the latter (values) and too short in the former (facts). Or, to sum up the critique in one word, such competing visions are – gasp – utopian (see for example Avery, 1999).

Rather than attempt to counter this critique, we embrace it and turn it on its head. For us, conventional agriculture is not utopian enough – or, to be more accurate, it is indeed utopic in its assumptions and dreams about how the world *ought* to look, but its proponents miss this fact because they've bought into the fictions of efficiency and science. Our choice of the term 'fiction' here is intentional. Karl Polanyi (1965) famously coined the term 'commodity fictions' to refer to artifacts – like property and money – that have been divorced from history and have thus been made to appear natural, self-evident, and objectively given. We argue the same holds for the so-called gods of science and efficiency: they are thoroughly utopian, in that they are grounded in conceptions of what the good life is and how we *ought* to achieve it.

Utopia here implies working for some future state of the world based on certain, deeply held assumptions. The primary arguments from large-scale conventional food systems actors revolve around these arguments of technology and efficiency that assumes unlimited resources and unlimited growth potential (in yield and profit). Van der Ploeg (2009, p19) refers to these assumptions as virtual realities of farming, where it is assumed that yields and profits will increase, anomalies arise when those increases do not occur. On the other side we locate those that argue for medium and small-scale agriculture. Within this camp include those who focus on cultural ways of farming and traditional food production, who mount their positions with the assumption that local, traditional, less mechanical and less chemically-dependent farming is better. Here the focus is on food as a medium for ideal communities without the involvement of large corporate holdings. This Arcadian utopia assumes an Edenic relationship to nature and provides a counterweight to the conventional utopian assumption of efficiency. In both cases, proponents' arguments come from somewhere.

It's time we began thinking about food production in these terms – that is, in terms of these background assumptions (those unsaid *oughts*). This changes the nature of the food debate entirely. Rather than, for example, proponents of organic agriculture scientifically proving that their system is sufficiently 'efficient' to compete with the conventional model, the debate instead takes a step back and deals first with those underlying assumptions. This brings to the fore such questions as: 'what does it mean to talk about "food security?"' and 'what constitutes "efficiency" when it pertains to food production?' Even a question like 'what is "food"?' needs to be fleshed out, for, ultimately, competing visions of food production rest in part upon competing visions of what *ought* to constitute as 'food'; Michael Pollan (2008, p7), for instance, famously argues that conventionally-produced food is not really 'food' at all.

Alas, space is limited. A book chapter only allows for so much conceptual unpacking. Eventually, categories have to be erected for the sake of an analysis of how these utopic visions differ. We would like to begin this discussion by offering the following premise: based upon our extensive shared experiences conversing with individuals about food and food security, it seems that the aforementioned competing utopian visions about food production (and, by extension, society) can generally be placed within one of the following two categories. Among proponents of conventional agriculture, food security is frequently reduced to matters of *quantity*, which is another way of saying it is about maximizing the production of calories, protein, vitamins, and the like. Critics of conventional agriculture, conversely, seem to be working from a more *qualitatively* oriented definition, which gives more room for questions about *how* food is raised, processed, and prepared (for instance, the cultural and social significance of food is lost when 'food' becomes an overarching term for a carrier of calories, carbohydrates, and protein).

Take the famous 2002 incident when the US sent significant quantities of food aid, in the form of whole kernel corn, to southern Africa. Soon thereafter it became known that the aid contained genetically modified organisms (GMOs), though the recipients had not been notified before the shipments were sent (see, for example, Clapp, 2004). Many in the developed North were outraged that any country would refuse food when its people were starving; in their minds whether the aid shipment contained GMOs or not was immaterial. Yet this misses the competing utopic visions about what food security *is*. If the debate were *only* about calories than perhaps such outrage would be justified. But that's not what the debate was about for everyone involved. Writing on the politics of famine, Jacques Ellul (1990, p53) explains:

> We must not think that people who are the victims of famine will eat anything. Western people might, since they no longer have any beliefs or traditions or sense of the sacred. But not others. We have thus to destroy the whole social structure, for food is one of the structures of society.

The purpose of this chapter is not to resolve the debate about what global food security 'ought' to mean. Our goal is more modest. We hope to highlight that conventional agriculture proponents and critics are both equally utopian in their visions about food security. Once we recognize this fundamental fact, we can begin to discuss and debate the real rub lying at the centre of most debates about food production, which we analytically distinguish as being between quantity and quality.

## Food security through the lens of quantity

Driving the interstate through 'farm country' in the US takes you past the inevitable billboard: 'One US farmer feeds 145 people'. At first we are taken aback by the sheer efficiency of American agriculture and in the spirit of advertising the billboard has worked. But does each farmer actually feed 145 people? How is that calculated? The calculations surely assume certain caloric intake levels and production levels that simple division then yields a neat and tidy quantification of the efficiency of the American farmer. While the number of people fed has continued to increase, are we any closer to achieving global food security than we were a couple of decades ago? The global food crisis of 2008 along with food aid in the form of food stamps, food banks and food drops suggest, sadly, that we are not.

A quantitative approach to food security assumes that if we simply produce enough, the market will ensure that everyone is sufficiently fed and that profits are earned from a job well done. Simple. We can limit labour input and improve efficiency (often the same thing) with increased use of machinery, GPS technology to best plot the crops, utilizing the latest data on soil quality, and with new fertilizer systems especially calibrated for the area. Research is then oriented to increasing yield to ensure that production can keep up with the rate of population growth. Gallantly, the US and other 'first world' producers take on the burden of that production – the noble providers. This provision comes in the form of caloric production through yield increases and economies of scale related to production that maximize water efficiency, distribution networks *and* calorie production. But is that what really happens?

### *Calorie production*

Let's examine the calorie argument first. Food security is often reduced to caloric production – if we produce enough food then hunger will be a thing of the past. It is an equation missing a few variables. Let's take, for example, the green revolution beginning in the 1960s. Miracle rice, so-dubbed for its miraculous qualities to increase yield, reduced the need for inputs by reducing plant size and shifting much of the energy into edible parts of the rice. The related elimination of biodiversity; from thousands of varieties of rice to just a few, led to a deficiency in Vitamin A in the rice produced. The recent technological solution is genetically engineered 'golden rice', with extra beta-carotene that can be converted into vitamin A in the body. Despite the fact that golden rice will only provide one per cent of the necessary daily intake of vitamin A, this is not part of the larger narrative – only the fact that a technological solution was found is important, regardless of how little it actually contributes to a solution (Shiva, 2000). Indicative of a quantity mindset – where a Vitamin A deficiency can be unproblematically resolved with simply more Vitamin A – golden rice ignores the fact that this vitamin is fat-soluble. In less affluent countries, however, levels of dietary fat are often insufficient. And simply increasing a population's daily intake of rice, by itself, contributes none of this life-sustaining fat to their diet.

The green revolution was fuelled in part by the development of a technological utopianism unique in the course of the twentieth century in scale. Technological utopianism at its heart advocates an 'if we can, we must' belief in technological salvation (Stivers, 1994). Applied to agriculture, it marries perfectly with a quantitative approach that believes (for it is on faith that it proceeds) that when we produce enough, there will be enough. And depending on the metrics involved, the world already produces enough food in caloric terms to feed the existing population and yet over one billion people are classified as undernourished.

## Growth assumptions

The calorie-based argument assumes yield improvements on a year-to-year basis, in the same way that corporations anticipate annual profit growths for investors (the only people who matter). Yet, while certain parts of the world could attest to gains in yield due to twentieth century breeding, recent genetic engineering (GE) techniques have done little to continue to push up those yields (Gurian-Sherman, 2009). In fact, GE techniques have pushed yields backwards, what is known as yield drag. Furthermore, the amount of arable land is decreasing. In wealthy countries, former farmland is turned into 'lifestyle blocks' or housing tracts while new land brought into production remains of dubious value. At the same time, much of the world grain stocks now go to the latest quick fix for oil dependence: biofuels. While brilliant on the face of it, in the end biofuels demand more oil to produce than the end product can give back to the system.[1] As food goes increasingly into making everything but food – food as livestock feed, food as fuel, and so on – the billboard described at the beginning of this section stating how many people a farmer feeds must be viewed with some doubt.

## Global land-grab

Similarly, increases in yields will do nothing if all the best land in the developing world is owned by outside interests. Oil-rich and land- and water-poor countries (Saudi Arabia, Bahrain, Kuwait, Oman, Qatar, and the United Arab Emirates) are investing heavily in many parts of Africa, Asia, and Eastern and Central Europe. Saudi Arabia, for example, owns massive wheat fields in Sudan. But the 'grabbing' is not limited to the Arabian peninsula. While China is buying up large tracks of land in Africa to produce biofuels and food (Spieldoch, 2009), South Korean company Daewoo tried to lease almost half of Madagascar's arable land for 99 years, with virtually no taxes or other benefits flowing back to Madagascar (The Economist, 2009). The public in Madagascar rose up in protest, which not only stopped the deal, but also led to the eventual overthrow of the government.

Attacking global hunger quantitatively reminds us of the war on terror: a war on the symptoms rather than the cause. If you had a headache caused by a brain tumour and your doctor prescribed acetaminophen instead of chemotherapy, radiation and/or surgery, you would change doctors. When food security is chopped into substitutable bits, solutions to global hunger take on this bit-like quality, where vitamin A is understood as a 'solution' even though vitamin A is a fat-soluble vitamin and rice a fat-less grain. The bits include yields per hectare, ratio of pounds of feed to live-weight gain, and ratio of seed-biomass to shoot-biomass. These are not unimportant but they do not describe the world *in toto* either, especially as far as global hunger is concerned. A recent study estimates that over 23 per cent of the world's population is overweight and an additional 10 per cent obese (Kelly *et al.*,

2008). Juxtapose this to the fact that today over half of the world's population is defined as 'hungry' (Dybas, 2009, p646). In light of fact like this, it's hard for us to believe that there's not enough to go around. Food just needs to be allocated differently. Increases in yields and improved conversion ratios do nothing in themselves to improve food access and reduce global poverty, the root causes of global hunger (Sen, 1981).

In decaying urban areas of once prosperous cities, not only is there a population unable to access any food, but the kind of 'food' also comes into question. Of course we produce a lot of food – but what kind of food is it? Is it food? Is a Twinkie food? For some, the only food options within a reasonable geographic proximity come from 'convenience' stores. The sarcastic quotation marks around convenience are intentional, for convenience food translates to increased rates of obesity, heart disease and diabetes. Wallinga (2010) connects the US 'cheap food' policies to the increasing rates of childhood obesity due to structural issues that reward processors who produce fatty, sugary 'food' with minimal nutritional value. And here too quantitative views of food security fail us. Advances in breeding and animal husbandry do nothing to break this addiction to nutritionally empty 'foods', except to make them even cheaper and more prolific.

## Food security through the lens of quality

Now that we've identified that the quantitative approach to food cannot (and, honestly, will not) adequately address today's food problems, we turn our attention to the qualitative approach – those ideas that recognize the cultural importance of food and how it is produced, as opposed to just having enough. These quality concerns tend to focus more on the forest than the trees; on all the stuff that makes food socially, culturally, and politically meaningful. Often this focus involves a turn toward the local, but 'quality' and 'local' are not equivalent. A country of origin label, for example, relies heavily on quality imagery without being local in the traditional sense (the last time Michael had New Zealand lamb was in California).

To be clear, we fully support the goals of movements like Slow Food, local food, and protected designation of origin (PDO) schemes. Through food, things like culture, community and identity are created, enacted, and reinforced. We also believe food needs to be produced in socially, environmentally, and economically just (it needs to be affordable) ways. How do we produce food that respects community and the environment but is also economically efficient? Let's be honest: it's a luxury these days to be a conscientious consumer; last time we checked, free range chickens and fresh local produce cost more than the quantitatively orientated food provided by the conventional food system. Admittedly, these 'costs' are in a significant degree the result of bad accounting: an artifact of externalities that others will be paying for generations. But price is price when you're living pay cheque to pay cheque.

## *Organics, permaculture and agroecology*

Organic farming (and the related agricultural methods) provides the obvious counterweight to quantity arguments (Rosin and Campbell, 2009). In many forms of organic, biodynamic, agroecological and permacultural production the focus begins with nutritious food grown in ways that respect the land, the soil and the people that depend on those resources for food and livelihoods. While subsistence – or having enough food to live healthily – remains an issue, the core focus is on the quality of the food and the production style. The different branches of the quality movement – such as vegan, vegetarianism, humane meat production, preservation of the peasantry, prevention of deforestation, water issues, etc. – then give those understandings of 'quality' and 'style' their own meanings. But the concern here is to not essentialize the argument of quality to an assumption of moral superiority, an issue we will return to below.

## *Slow Food*

We classify the Slow Food movement as typifying what we call quality issues surrounding food security. While famous for protesting at a McDonald's at the base of the Spanish Steps in Rome, the complaint was not against McDonald's *per se* but against the symbolic (food) violence perpetuated by a food system enamoured by the logic of quantity. Slow food celebrates diversity and local knowledge. It highlights cultural, regional, and national foods and encourages a sharing across borders, both real and imagined. Its emphasis on the quality of the food, and the nurturing of social relationships during food preparation and consumption that help sustain cultural knowledge and practices, make Slow Food a quality movement.

Yet being 'slow' for the sake of slowness doesn't feed people either. Feeding a soul is fine when the body is nourished. We still need to produce food – a lot of it – if we want to feed the world's growing population. Wendell Berry, for example, worries that, given the West's collective detachment from the land and food production, we might not have the embodied and explicit knowledge any more to raise sufficient food via alternative methods. We doubt we'll see a significant back-to-the-land-movement in developed countries, at least as traditionally conceived. Nor do we necessarily think that such a thing would automatically improve food security, recognizing that while people may be moving back to the land they are not moving back in droves to agriculture, leading to rural gentrification (Phillips 1993).

Moreover, we know first-hand just how difficult it is to fix roots to any particular place. Michael has lived in Iowa, Illinois, the Netherlands, Washington state, and Colorado in a matter of just a couple of years. Paul too has found it difficult to stay in just one place, moving from Missouri, Illinois, Colorado, and New Zealand over the last eight years. The point of this brief author biography is to emphasize the rarity, at least in the increasingly mobile classes, of a connectedness to place. How does one practice slowly forming quality relationships to people, places and things when they're constantly on the move? We don't have an answer to this question. It's more rhetorical; to remind ourselves of those who do not (cannot) remain in one place for an extended period of time, or of those who might live in a place lacking a meaningful food biography (what about those living in, say, Las Vegas, which, as the world's cities go, was born yesterday?).

### Home and urban gardening

Home gardening (relatively ubiquitous in New Zealand) and urban gardens (being rediscovered in the US following the recent downturn in the economy) may help to assuage Wendell Berry's fears of not enough knowledge to grow enough food. But it's always easier to discard years of cultural knowledge than it is to retrieve it (Carlsson, 2008). Take the case of Detroit. Detroit is, in many respects, the birth-place of the automobile (or at least of the mass-produced automobile). And as the auto industry has struggled, so has the city. The city today is a hollow replica of its former self. The city's subsequent loss of population and its failing infrastructure lead to possibility, though. Dowie (2009) points to multiple efforts to re-establish self-sufficient food production in Detroit. Creative destruction, as it pertains to local food production. As Dowie explains:

> One obvious solution is to grow their own, and the urban backyard garden boom that is sweeping the nation has caught hold in Detroit, particularly in neighbourhoods recently settled by immigrants from agrarian cultures of Laos and Bangladesh, who are almost certain to become major players in an agrarian Detroit. Add to that the five hundred or so twenty-by-twenty-foot community plots and a handful of three- to ten-acre farms cultured by church and non-profit groups, and during its four-month growing season, Detroit is producing somewhere between 10 and 15 per cent of its food supply inside city limits – more than most American cities, but nowhere near enough to allay the food desert problem.
>
> Dowie (2009)

Just as Detroit symbolizes the failure of urbanism to some, it also acts as a demonstration ground for new possibilities. Instead of the demonstration plots of Monsanto or Michigan State, we have new demonstration plots sponsored by smaller companies, Wal-Mart, neighbourhood organizations and venture capital-ists. What were once considered guerilla gardens are being recast as spaces of entrepreneurialism and community, defined in multiple ways.

The unifying theme of these examples – organics, Slow Food, and home and urban gardening – of the food quality lens focuses on a positive vision of what food can, and should, look like. Yet, as with all utopic visions, they too are prem-ised upon certain *oughts*. This is not meant to minimize their argumentative affect. Any and all positions concerning how food security *should* be achieved are premised upon value judgments. We just ask that these value judgments be brought to the discursive surface; that they are not essentialized. The idea, then, of privileging the local suffers some of the same myopia of the quantitative view – which in this case is known as the 'local trap'. The local trap:

> refers to the tendency of food activists and researchers to assume something inherent about the local scale. The local is assumed to be desirable; it is preferred a priori to larger scales. What is desired varies and can include ecological sustainability, social justice, democracy, better nutrition, and food security, freshness, and quality ... the local trap is the assumption that local is inherently good.
>
> Born and Purcell, (2006, p195)

At multiple levels, the assumption of scale is equated with outcome. For many food activists, global equates with 'bad' and local equates with 'good'. Essentializing the local in this manner prevents any critical discourse from turning back onto these assumptions (which are taken to be objectivity given). As Polanyi (1965) convincingly argues, markets have always (at least as far as recorded history goes) been a part of people's lives, regardless of how 'global' or 'local' they looked. In nineteenth century Canada, for example, general stores carried a variety of imported goods. Even in rural Canada in the early 1800s, as one historian noted, 'It is hard, for example, to picture rural isolation from markets if farm families regularly visited country stores to purchase routinely consumed goods like tea and tobacco' (McCalla, 2005, p150). Similarly, we're hesitant to ascribe 'goodness' or 'badness' to production scale – we've both seen small farms managed terribly (and whose products were loaded into the back of a gas-guzzling pickup truck and driven a hundred miles to a large farmers' market) and large farms managed brilliantly (and whose products were consumed within a 15 mile radius). Besides, if it's true that the era of a countryside laced with small farms is over – indeed, as Laura DeLind (2010, p4) points out, even those with the resources to grow food often don't (these she calls 'lazy locavors') – we'll need farms of significant scale to (quantitatively and qualitatively) feed us.

This argument is not meant to derail those calls for carrying out food production and provisioning differently, particularly those interested in shortening the space between producer face and consumer fork. It seems to us what is needed, rather, is greater reflexivity all around, by quality and quantity proponents alike. Before concluding we would like to offer some thoughts about what this balancing act might look like through the example of vertical farming.

## Getting the discussion going: vertical farming

Rarely do you hear those rallying against conventional food, rallying behind technology. Where does one find culture or community in 'cold' technology? Technology is about instrumental rationality, about quantity. Yet, if we can admit that romanticized views of the countryside are just that, romanticized views of an invented past – even Wendell Berry recognizes the difficulties of repopulating the countryside with farmers – then perhaps we can get over this aversion many food activists have towards capital-intensive systems of food provision. On the other 'side', vertical farms represent such a radical departure from how agriculture has been done (I doubt Earl Butz had this in mind when he was talking about 'getting big or getting out') that proponents of quantity, likewise, give little thought to it.

Vertical farms have fired the imagination of some, however, especially those concerned about food security in crowded urban centres (see, for example, Despommier, 2010). Vertical farming may not sit well with those holding romanticized images of agriculture, for it jumbles up old utopian views that place food production, rurality, and nature in the same categorical box. Instead of worrying about the geographic spread of arable land, why not develop farms in the same direction as most plants go: up.

Vertical farms bring food (and 'farms') to inner city populations since, if we are to be honest with ourselves, it is unlikely that the reverse is going to happen. Vertical farms could bring jobs to these communities, while also preserving local, cultural cuisine by growing crops specifically for the community (Slow Food using state-of-the-art technology!). Vertical farms may be more emblematic of the direction food production will have to go as one part of reconstructing an effective food system that pragmatically combines the qualitative and the quantitative. At the same time, this model of food production might work really well in places like New York City, Beijing, and Hong Kong, where affordable agricultural space is limited.

Vertical farms offer a philosophical topic to bridge the qualitative and the quantitative assumptions embedded in these contrasting utopian visions of food production. Furthermore, the very idea of vertical farms – agriculture in the city – helps broker-related dichotomous stumbling blocks to open discussion related to food – the urban and rural, the local and the global, jobs versus the environment and people versus nature. Vertical farms and the projects related to their establishment, operation and integration into place, provide room to discuss not just food, but what societies can, and should, look like.

## Discussion and conclusion

A dominant frame used to talk about food security is how 'cheap' it is. It is a powerful rhetorical tool for proponents of the conventional food system; pointing out that their food is cheap (and they're working hard to make it cheaper), while the alternative is not (Carolan, 2011). We alluded to this above. Though an artifact of bad accounting, there's no denying that, on the whole, conventional food, in terms of its price at the grocery store, is cheaper than food made and produced with a greater eye toward ecological and social sustainability. We prefer to frame the debate as being one about expense. We want inexpensive food, not food that is cheap. Conventional agriculture might produce cheap food but it does so only with great expense, to culture, biodiversity, human and environmental health – even global warming can be strongly tied to Big Ag's myopic focus (see, for example, FAO, 2006, p101). Understanding food security in a way that balances both quantity and quality will allow us to talk about producing food in a way that produces fewer expenses.

This myopic gaze is also reflected in many food aid policies. For many countries, since 'food' has been reduced to 'calories', food aid could be renamed 'calorie aid', which is basically what it is. When countries like the US set out to provide food aid to countries ravaged by famine, they send calories – a band-aid solution at best to a systemic problem. For example, the USAID (United States Agency for International Development) food aid budget in 2005 was US$1.6 billion. Of that, only 40 per cent (US$654 million) went towards paying for food. The rest was spent on overland transportation (US$141 million), ocean shipping (US$341 million), transportation and storage in the destination countries (US$410 million), and administrative costs (US$81 million) (Dugger, 2005). Perhaps this money could be spent more optimally. The world's top ten crop-science companies spend approximately US$3 billion a year on biotech research (Karapinar and Temmerman, 2008, p192). Compare this to the US$30 million spent annually by CGIAR (Consultative Group on International Agricultural Research) on biotech research aimed at benefiting farmers in the developing world (about seven per cent of its annual budget). If the US conceived of 'food aid' a little differently, funds equal to more than half of what the world's top ten crop-science companies spend on biotech research would be available annually to breeding projects aimed at the needs of farmers in the developing world. In summary, policies directed at improving global food security, when done with an eye towards both quantity and quality, need to balance short-term realities (recognizing that people are starving and need to eat now) with long-term visions (those people will need to continue to eat in the future).

To put it simply: 'Cheap food causes hunger' (Wise, 2010). Economic policies supported by the World Bank and IMF (International Monetary Fund) treat food and food production as just another form of manufacturing. But food is different. People do not need cars or other consumables to sustain life and carry on tradition. But the changes in Mexico's corn production following NAFTA (North American Free Trade Agreement) illustrate not a tragedy of the commons, but a tragedy of a faith in markets. We are taught in our 'introduction to economics' classes that comparative advantage speaks to agro-ecological conditions, to something inherent to a region or country that gives them an advantage over others (Kona coffee, from the 'big island' of the Hawaiian Islands, comes to mind here). But more often than not, comparative advantage is an artifact of market distortions, of subsidies, export subsidies, tariffs, and government-funded infrastructure build up. Case in point: Mexico. Having produced corn for millennia, how else could Mexico NOT have a comparative advantage over the US in corn production?

While food needs to remain cheap either via government or international regulations, changes in food habits are also constrained by culture. A few years ago, a star professional baseball player switched to a vegetarian diet. After struggling at the beginning of the season, many sportswriters and former players attributed the lack of performance to the dietary switch, as well as conflating vegetarianism with femininity. Meat consumption around the world is associated with progress – intellectual and economic – and masculinity. To be able to eat meat is to have arrived. And just offering alternatives like Slow Food, or even healthier options at McDonald's, is not going to be enough to challenge the rising public health issues related to bad food policy. Structural problems require structural changes. And that's hard.

But, like the utopian assumptions that underpin arguments across the food spectrum, the cultural superiority of meat is a myth. Just like the inherent problems challenging the corporate interests of the food pyramid in the US (Nestle 2003), changing ingrained cultural assumptions is difficult (thus the opposition to Lester Brown and Pollan and others who advocate eating lower on the food chain). But myths are powerful and they often prevent dialogue. Let's use the fishing proverb. It helps us examine the utopian assumptions of both sides. The quantity people argue that if people can fish for bigger fish, genetically modified with more calories, more people will be fed. The quality people want to encourage respect for fish stocks and tradition. How do we reconcile a respectful, yet pragmatic lesson from a parable with different assumptions about the world?

Looking at these competing utopic visions we can say at least that they can both learn a little from the other, which is good because it means that there is room for dialogue and mutual learning. Unfortunately, the debate between quantity and quality is reduced to a crass caricaturization of one another. We can almost see the editorial cartoon featuring tiny, exotic vegetables parading with placards asking, 'WHAT ABOUT THE CHILDREN?' about to be stepped on by a giant corn cob labelled Big Ag. The root problem lies in a false distinction between having enough food and what kind of food. Essentially, we can reduce the problem to a Mannheimian sociology of knowledge in which both sides have key and important arguments to make, but each employs a different language rooted in their first assumptions of what ought to happen (Mannheim, 1985).

Real steps forward require demystifying the utopian superiority enjoyed by proponents of both quantity and quality arguments. Moreover, we can't help but see the irony in a movement that calls for a food system that rests upon respecting local values, cultures, biographies, and agro-ecological conditions that has gravitated toward a figurehead telling an entire nation how to eat (see Pollan, 2009) and how the food system *ought* to be (see Pollan, 2008). At the same time, we have a system that places into key decision-making positions, at seemingly every opportunity, someone from the quantity side. For instance, the appointment of a former Iowa Governor as head of the US Dept. of Agriculture has won little support from the qualitative side. It has become a debate of extremes.

Of course, the US and Europe are key actors here, and results are related to the oft-ignored importance of the Farm Bill and trade negotiations often inaccessible to public oversight. Make no mistake: cheap and less nutritious food is profitable.

Although the idea of a nanny state as iron chef setting the menu reeks of totalitarianism, policies directed at ensuring access to more nutritious food would seem to be in the best interest of states, companies and individuals. A multi-pronged effort to ensure nutritious food access would involve regulation of food production; a re-evaluation of national and international food policies that favour production for productions' sake. Food is not a car. Food is different and we need to talk about it differently. We need to reorganize transportation systems and zoning to ensure accessibility to stores, farms, and food. Any solutions to issues of global food security will require a political will that, at present, only seems capable of coalescing around war (see Brown, Plan B 4.0, 2010). But, just like climate change, food security demands creative solutions operating at multiple levels. There are, unarguably, ways to provide options. And for those of us with a choice, these involve working in our own small ways, while working collectively in big ways. Sometimes that is all we can do – even if it never feels enough.

## Note

1    There is also the dichotomy of corporate biofuel, which McMichael calls agrofuel, versus the open-source, veggie oil 'movement', which utilizes recycled vegetable oil from restaurant fryers (see Carlsson, 2008; Lawrence, Lyons and Wallington, 2010).

## References

Avery, D. (1999) 'The Fallacy of Organic Utopia', in J. Morris and R. Bate. (eds) *Fearing Food: Risk, Health, and Environment*, Butterworth-Heinemann, Boston

Born, B. and Purcell, M. (2006) 'Avoiding the Local Trap: Scale and Food Systems in Planning Research', *Journal of Planning Education and Research*, vol 26, pp195–207

Brown, L.R. (2010) *Plan B 4.0: Mobilising to Save Civilization*, W. W. Norton and Company, New York

Carlsson, C. (2008) *Nowtopia*, AK Press, Edinburgh, Scotland

Carolan, M. (2011) *The Real Cost of Cheap Food*, Earthscan, London

Clapp, J. (2004) WTO Agricultural Trade Battles and Food Aid, *Third World Quarterly* vol 25, no 8, pp1439–52

DeLind, L. (2010) 'Are Local Food and the Local Food Movement Taking Us Where We Want to Go? Or Are We Hitching Our Wagons to the Wrong Stars', *Agriculture and Human Values,* Online first (in press)

Despommier, D. (2010) *The Vertical Farm: Feeding the World in the 21st Century*, Thomas Dunne Books, New York

Dowie, M. (2009) 'Food among the ruings', *Guernica*, August, available at www.guernicamag. com/spotlight/1182/food_among_the_ruins/, last accessed September 2011

Dugger, C. (2005) 'Africa Food for Africa's Starving is Road Blocked in Congress', New York Times October 12, www.nytimes.com/2005/10/12/international/africa/12memo. html?ex=1286769   600&en=0de1afa6dd7990e7&ei=5090&partner=rssuserland&emc =rss, accessed 19 February 2010

Dybas, C. (2009) Report from the 2009 AIBS Annual Meeting, 'Ensuring a Food Supply in a World that's Hot, Packed, and Starving', *BioScience* vol 59, no 8, pp640–646

Ellul, J. (1990) *The Technological Bluff*, Eerdmans, Grand Rapids, MI

FAO (Food and Agriculture Oranization) (2006) *Livestock's Long Shadow: Environmental Issues and Options*, FAO, Rome, Italy

Gurian-Sherman, D. (2009) 'Failure to Yield: Evaluating the Performance of Genetically Engineered Crops', Union of Concerned Scientists, Cambridge, MA, www.ucsusa.org/food_and_agriculture/science_and_impacts/science/failure-to-yield.html

Karapinar, B. and Temmerman, M. (2008) 'Benefiting from Biotechnology: Pro-poor IPRs and Public-Private Partnerships', *Biotechnology Law Report,* vol 27, no 3 pp189–202

Kelly, T., Wang, W., Chen, C.S., Reynolds, K. and He. J. (2008) 'Global Burden of Obesity in 2005 and Projections to 2030', *International Journal of Obesity,* vol 32, pp1431–1437

Lawrence, G., Lyons, K. and Wallington T. (Eds.) (2010) *Food Security, Nutrition and Sustainability*, Earthscan, London

McCalla, D. (2005) 'A World Without Chocolate: Grocery Purchases at Some Upper Canadian Country Stores, 1808–1861', *Agricultural History*, vol 79, no 2, pp147–172

Mannheim, K. (1985) *Ideology and Utopia*. Harcourt Brace Jovanovich, New Yo

Nestle, M. (2003) *Food Politics: How the Food Industry Influences Nutrition and Health*, University of California Press, Los Angeles, CA

Phillips, M. (1993) 'Rural Gentrification and the Process of Class Colonization', *Journal of Rural Studies,* vol 9, no 2, pp123–140

Polanyi, K. (1965) *The Great Transformation: The Political Economic Origins of our Time*, Beacon Press, Boston, MA

Pollan, M. (2008) *In Defense of Food: An Eater's Manifesto*, Penguin Group, New York

Pollan, M. (2009) *Food Rules: An Eater's Manual*, Penguin, New York

Rosin, C. and Campbell, H. (2009) 'Shifting moralities of sustainable agriculture: evidence of organification in New Zealand'. Paper presented at *Agri-Food XVI: Annual Meeting of the Agri-Food Research Network*, Auckland, New Zealand, November 2009.

Sen, A. (1981) *Poverty and Famines*, Oxford University Press, New York

Shiva, V. (2000/1) 'World in a Grain of Rice', *The Ecologist* (Dec/Jan), www.mindfully.org/GE/Rice-Vandana-Shiva.htm

Spieldoch, A. (2009) 'Global Land Grab', *Foreign Policy in Focus*, June 18, Washington, DC, www.fpif.org/articles/global_land_grab, accessed 22 February 2010

Stivers, R. (1994) *The Culture of Cynicism* Blackwell, Hoboken, NJ

*The Economist* (2009) 'Buying Farmland Abroad: The Scramble for Land in Africa and Asia', *The Economist*, May, 21, www.economist.com/world/international/displaystory.cfm?story_id=13692889, accessed 22 February 2010

Van der Ploeg, J. D. (2009) *The New Peasantries*, Earthscan, London, UK

Wallinga, D. (2010) 'Agricultural Policy and Childhood Obesity: A Food Systems and Public Health Commentary', *Health Affairs*, vol 29, no 3, pp405–410, (http://content.healthaffairs.org/cgi/content/full/29/3/405)

Wise, T. (2010) 'The True Cost of Cheap Food', *Resurgence*, vol 259 (March/April), www.ase.tufts.edu/gdae/Pubs/rp/TWG20ResurgenceMar10.pdf

# Part 2

# THE CONDITION OF
# NEOLIBERAL AGRICULTURE

# 9

# CLIMATE CHANGE AND THE RESILIENCE OF COMMODITY FOOD PRODUCTION IN AUSTRALIA

*Geoffrey Lawrence, Carol Richards, Ian Gray and Naomi Hansar*

## Introduction

The current view of Australian state and national governments about the effects of climate change on agriculture is that farmers – through the adoption of mitigation and adaptation strategies – will remain resilient, and agricultural production will continue to expand. The assumption is that neoliberalism will provide the best 'free market' options for climate change mitigation and adaptation in farming. In contrast, we argue that neoliberalism will increase the move towards productivist ('high-tech') agriculture – the very system that has caused major environmental damage to the Australian continent. High-tech farming is highly dependent upon access to water and fossil fuels, both of which would appear to be the main limits to production in future decades. Productivist agriculture is a system highly reliant upon fertilizers and fuels that are derived from the petrochemical industry, and are currently increasing in cost as the price of oil increases.

After examining the structure of food production in neoliberalist Australian agriculture, this chapter outlines the climate change challenges faced by Australian producers. It investigates the most likely climate change scenario facing farmers in the Murray-Darling Basin (Australia's 'food bowl') and finishes by outlining the adaptation options for farmers under the current neoliberal governance regime. It concludes by acknowledging that, while productivism has been a key to securing Australia's domestic food supply and export-oriented food production in the past, it is now a barrier to more enlightened, environmentally sustainable approaches to food production for the future.

We seem to be at a particularly interesting moment in the political history of productivist agriculture in Australia, especially as concerns of environmental sustainability, peak oil, climate change, and global food security have piqued public interest. For the first time in Australia, there appears to be a serious attempt by government to address one of the more pressing issues linked to productivist agriculture – water use. Amid the debate, many analysts blame the government for having built the dams and channels – and for having given away the licences – that have facilitated the burgeoning of farm output in areas like the Murray-Darling Basin (MDB). But to 'blame' government is a mistake. As we will argue below, the over-allocation of water was just one element in the creation of an export-oriented, technologically 'advanced' form of farming which is now found wanting.

Despite a fast approaching environmental calamity for the MDB, together with other over-allocated waterways in Australia – there exists a strong local discourse that configures such areas not only as the 'food bowl' of Australia, but crucial to Australia's export economy. In the fall-out from the release of a 2010 government report on water allocation in the MDB, farmers held a prominent position in media discourses claiming that the government's cack-handed management of the issue would result in a demise in profitable farming, the inflation of food prices and a lack of supply of food to the masses in Australian capital cities.

The tensions between social, economic and environmental sustainability are persistently being played out over resource use in Australian agricultural and pastoral areas, such as the Murray Darling Basin. At the heart of this concern is a cautious, but emerging, challenge to the productivist paradigm that has been char-acteristic of Australia's post-war agricultural development. Given we now know that Australia's high-tech system of farming is failing its citizens and the environ-ment, we should hope the government will be even *more* interested in intervening to correct the market deficiencies that accompanied its earlier actions. This will mean creating a totally different trajectory for Australian agriculture.

## The evolution of the current structure of production

European farming methods and ideologies accompanied the settlement of the convicts of the First Fleet in Sydney in 1788. Indeed, one major aim of estab-lishing the colony was to remove the burden on Britain of housing and feeding its prisoners. While it was several decades before grains could be grown and harvested in sufficient quantities to provide a secure domestic supply, and later a supply for export, it was clear that sheep and cattle would do well on the pastures beyond the Great Dividing Range. Wool, wheat and sugar were convenient prod-ucts for transportation to Britain in the first six or seven decades of colonial occu-pation, later supplemented by frozen lamb and beef when refrigerated vessels began to leave Australian shores in the 1870s. In line with Britain's interests in securing food and fibre from its colonies, Australia developed an export-oriented agricultural system.

The reliance upon trade in agricultural products during the first 150 years of settlement gave rise to the expression that Australia 'rode on the sheep's back' (currently, it rides on the back of mining dump trucks). A settler colonial history of British legal systematics, property rights and sociopolitical arrangements, coupled with European-style farming methods and the exporting of food staples – fortified by the 1850s gold rush – produced strong economic growth. Federation of the Australian states in 1901 created national coherence in political decision-making and provided a firm foundation for nation-building activities. During the first seventy years of federation, protectionist policies and structures nurtured a widespread and robust system of family-farm agriculture which was accompanied by productivity-boosting state-based rural research and extension.

When Britain joined the European Economic Community in 1973, markets for many of Australia's agricultural exports collapsed. State and federal governments of the day reacted by providing substantial public funds for rural 'adjustment' to reduce market oversupply and to restructure farming along more efficient lines. In some cases this meant paying farmers to pull out their healthy fruit trees or grape vines, in others it meant giving farmers a 'golden handshake' to leave the land. For a number of policy analysts – particularly conservatively-trained Canberra-based agricultural economists in the Australian Bureau of Agriculture and Resource Economics (ABARE) – public 'handouts' were anathema to an emerging free-market ideology. To these bureaucrats, it was clear that the protectionist settings that had given some stability to farmers were now hindering capital accumulation. They argued strongly for the removal of regulatory barriers to trade and endorsed the importance of competition in agriculture. Exposure to 'market forces' was shorthand for embracing increasing returns to scale, fostering intensification and specialization, promoting the closer links of farming to corporate agribusiness, and removing government protection and subsidization from agriculture. Facing continuing cost-price (terms-of-trade) pressures, farmers struggled as they were progressively exposed to the fickle world marketplace. Yet, the farmers' own industry organization, the National Farmers' Federation (NFF), endorsed this move: it was told by ABARE that Australian farmers could reap billions of dollars if free trade settings were put in place nationally and globally. Since the 1980s, agricultural policy in Australia has been dominated by neoliberalism and the economic rationalist policy settings (privatization of state services, liberalization of state regulations, cutbacks in redistributive spending) which accompany that ideology (Gray and Lawrence, 2001).

Neoliberal policy settings combine a strong determination to have individuals take responsibility for their actions (individualization), along with a belief that free markets will – and ought to – sort those who are innovative and profit-driven from those doomed to competitive failure (marketization) (Lockie et al., 2005). Individualization and marketization are in direct opposition to planned government interventions: in fact, governmental 'interference' is viewed as standing in the way of progress. Neoliberal governance is based upon the extension of property rights, including the right of those with property to do what they wish with labour, capital – and, in the case of farmers – natural resources. In a productivist system this means expansion of capital and greater exploitation of the environment.

The form of agriculture that emerged from the 1970s and 80s under neoliberal policy settings was one that sought to give individual farmers the greatest freedom to maximize on-farm efficiency and to achieve the highest output. In competition with each other, and with overseas producers, they sought to increase output by harnessing the latest seed varieties, animal breeds, fertilizers, insecticides, veterinary chemicals, farm machinery, and by embracing mono-cultural practices (fence-to-fence planting of the same high-yielding grains) together with so-called 'factory farming' of chickens, pigs and cattle in concentrated animal feeding operations (CAFOs). Ever-increasing volumes of water for irrigated cropping were diverted from inland river systems, and native vegetation was cleared to make way for increased cropping and grazing. Finally, agribusiness firms grew strongly from the sale of inputs to farmers, as well as by transporting, packaging, processing and selling farm products to consumers. Success is measured by output: Australian agriculture currently feeds its domestic population of some 20 million people and exports sufficient food to feed another 40 million (PMSEIC, 2010, p15). What could be wrong here?

## Productivism and its discontents

Increasingly exposed to volatile 'free' markets and terms-of-trade declines, Australian farmers have eagerly embraced productivism as a means of expanding output and improving their competitiveness. They tend to *specialize* in one product; they *intensify* their operations, looking to utilize the latest agribusiness inputs that promise greater production efficiencies; and they attempt to gain returns to scale by expanding the *size* of their operations. The combination of these leads to *economic concentration* in agriculture – fewer farmers can produce ever-increasing volumes of food and fibre for the domestic market and for export (Gray and Lawrence, 2001; Argent, 2002).

Although the productivist pillars of specialization, intensification, farm size increase, and concentration have provided a solid foundation for the expansion of 'high tech' farming under neoliberalism, there have been a number of significant impacts. In fact, it could be argued that productivism has been implicated in one of the worst cases of collective environmental vandalism ever perpetrated on the Australian continent. The clearing of native vegetation for the expansion of farming has resulted in soil erosion, soil acidification and salinization, species decline and the release of carbon dioxide into the atmosphere (PMSEIC, 2010). Water used for irrigation has been extracted from fragile inland river systems – leading to serious ecosystem deterioration, including destruction of wetlands and species loss. The synthetic fertilizers employed in intensive farming have run into waterways, polluting streams and stimulating the growth of invasive weeds and poisonous algal blooms. Toxic agricultural pesticides have leached into soil and water systems, compromising ecosystem health and generating resistance in target pests. Increasing pest resistance places the farmer on a so-called 'technological treadmill' – having to use ever-more potent (and expensive) chemicals each year to control infestations. Finally, the whole system of synthetic fertilizers and pesticides and the fuelling of heavy agricultural machinery is one totally reliant upon the petrochemical industry. A key outcome of productivism has been the production of cheap foods; however, relatively low costs of foods are subsidized through an ever-depleting environment. With the cost of oil rising (and predicted to rise into the future as a consequence of 'peak oil'), food prices are also predicted to rise. The whole system of productivism can therefore be considered economically vulnerable and environmentally unsustainable (Gray and Lawrence, 2001).

## The challenge of climate change: Australia and the Murray-Darling Basin

Climate change adds another layer to the concerns about productivist agriculture in Australia. Predictions are that:

- Australia will be one of the most adversely affected nations of the world in relation to declines in agricultural production.
- Australian wheat, beef, dairy and sugar production are expected to decline by over 11 per cent (and potentially up to 60 per cent) by 2030, and by at least 15 per cent (and potentially up to 79 per cent) by 2050.
- Changing temperature and rainfall regimes will lead to more extreme weather events (floods, droughts, cyclones).
- While there will be moderate cropping increases in north-eastern Australia, there will be significant crop reductions in south-western and inland regions.
- There will be an increased spread of exotic weeds and native woody species.
- Pests such as the Queensland fruit fly and the cattle tick will increase their prominence in eastern Australia.
- Waterlogging, soil acidification, soil erosion and dryland salinity will increase.
- There will be salinization of waters used for irrigation, and coastal freshwater systems will become saline in areas affected by rising sea levels (ABARE, 2007; Gunasekera et al., 2007).

It has been estimated that if temperatures increase by more than two degrees, pasture growth will decline by over 30 per cent, with a subsequent reduction in livestock carrying capacity of approximately 40 per cent (Australian Government, 2010). In the face of these productivity-limiting predictions, there is no choice – according to ABARE – but for farmers to continue to 'maintain strong productivity growth' or leave the industry through farm 'adjustment' (see Gunasekera et al., 2007, p657). These are the same old prescriptions touted by the Bureau for decades – but they continue to fit nicely with the promotion of an industrial agriculture/agribusiness future for Australia. Adaptation measures such as the harnessing of new technologies – including genetically-modified organisms – are now being endorsed by the Bureau as key to the overall success of Australian farmers in dealing with climate change (Nossal et al., 2008). There is no discussion of alternative approaches to farming, and – for a Bureau with 'Resources' as part of its title – very little focus on resource protection and sustainability. In fact, during the decade of the Howard government ABARE, as a handmaiden to government neoliberalism, is accused of: overstating the costs to Australia of introducing carbon abatement measures; opposing state-based legislation aimed at preventing land clearing; and, until 2006, denying that climate change was occurring (Keane, 2008). So much for bureaucratic impartiality! Such is the faith that some institutions have in the market system that they are able to dismiss the risks associated with climate change by simply believing market forces will lead us to salvation. This is a statement of faith, rather than being an empirically-proven argument. That the market cannot properly cost 'externalities' – and most of the externalities are environmental costs that are unpaid by business – indicates the sleight of hand that is perpetuated by the proponents of neoliberalism. Let's recall Sir Nicholas Stern's comment on climate change: climate change, he argued, was the biggest market failure the world had ever seen. The greenhouse gas releases, and other forms of pollution, were never costed and built into the business models of private firms. Those firms pocketed the profits and socialized the costs of their activities – a process convenient for the firms, but disastrous for the environment and planet.

It is instructive to consider conditions in Australia's Murray-Darling Basin – see map in Figure 9.1. The Basin, which covers one seventh of the continent, currently produces 40 per cent of the gross value of Australia's agricultural production. Some 85 per cent of all agricultural irrigation in the nation takes place in the MDB, helping to produce about $9 billion worth of produce each year. Approximately 97 per cent of the Basin's wool, 80 per cent of its wheat and 50 per cent of its beef are exported. But soils in many parts of the Basin are thin, fragile and lack fertility. To achieve levels of production to sustain high-output farming, these soils receive heavy applications of nitrogen and phosphate. The climate is highly variable and there have been at least ten major (and 30 minor) droughts during the last century. Irrigation has, therefore, proven to be an essential ingredient in achieving high levels of agricultural production in the Basin.

A National Land and Water Audit in 2001, along with subsequent research, has established that:

- 40 per cent of the length of the river system is impaired (there is a marked reduction in the presence of macro-invertebrates).
- 95 per cent of the river system is despoiled (with high nutrient and sediment loads).
- Habitat condition is degraded throughout the Basin, and there has been a significant loss of riparian vegetation.
- Fish populations are deemed to be in the range 'very poor to extremely poor'.
- Wetland quality has been seriously reduced.
- Twenty mammal species have become extinct since 1900, and the iconic Murray Cod (Australia's largest freshwater fish) is in severe decline.
- A further 35 bird species and 16 mammals are currently listed as endangered.
- Climate change is expected to dramatically affect fish, mammal, bird and plant life (CSIRO, 2008; Department of Sustainability, Environment, Water

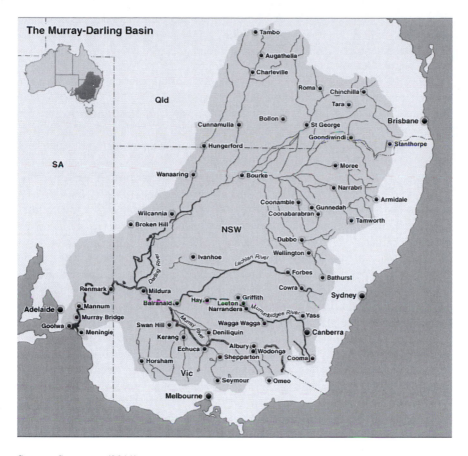

Source: Spraggon (2011)

*Figure 9.1* The Murray-Darling Basin

Population and Communities, 2010a).

As a result of these concerns, and others over water quality, the 11,000 kilometers of river systems in the Basin had been described – in the decade preceding the audit – as the 'world's longest sewer'. Water quality is affected by turbidity, industrial water and sewage, agrochemicals, and salts – in the Basin, over 1.3 million tonnes of salt are carried through the system, leading to readings well above World Health Organization recommended drinking-water levels (see Beale and Fray, 1990). In 2007 and 2008, researchers found that the extent of water use in the Basin was reducing the average stream flow to the mouth of the Murray river by over 60 per cent, with the river now ceasing to flow to the ocean 40 per cent of the time (in comparison, this was estimated at only 1 per cent prior to European settlement) (CSIRO, 2008; Department of Sustainability, Environment, Water, Population and Communities, 2010a).

State and federal governments have struggled to find solutions to the problem of environmental degradation in the Basin. The reason they have struggled is that they have, themselves, been a major part of the problem. State governments, for example, have provided leases for irrigation well beyond the capacity of the system to provide that water. Over the last 100 years, water use in the MDB has increased five-fold, partly because state governments have been issuing licences for irrigation as a means of stimulating farming output. But they have often done so without reference to similar efforts by other states with jurisdiction in the MDB. They have put their own farmers, and their profits, ahead of Basin-wide environmental concerns. In fact, an earlier MDB agreement was something of a farce: The MDB Ministerial Council had very limited powers, and the states were not prepared to concede power to an authority that might proscribe limits to productivist agricultural expansion; there was little interaction between natural resource management agencies of the various states in the Basin; and, the pro-development State of Queensland – which covers 20 per cent of the Basin and is the source of some 25 per cent of water entering the Basin – was not even a signatory to the MDB agreement until 1996 (see Crabb, 1988; Lawrence and Vanclay, 1992; Murray-Darling Basin Authority, 2010). Indeed, it was not until 1994 that the Council of Australian Governments formally recognized that 'the environment' was a legitimate user of water (Council of Australian Governments, 1994). Before that time, water was viewed almost exclusively as an input to regional agricultural and industrial production and for the domestic water needs of regional settlements.

The environmental problems of the MDB came to a head at the height of the most recent drought from 1997 to 2008, a time at which inflows into the Murray River were at historically low levels. The famous River Red Gums on the Basin's floodplains were severely water stressed and flows into the lakes and wetlands were a mere trickle, threatening biodiversity and drawing attention to the extent of irrigation being undertaken. In 2009–10 an agreement was signed between the prime minister and the premiers of New South Wales, Victoria and South Australia on the 'sustainable' distribution of water in the southern Murray-Darling Basin for this period. The federal government also launched its *Water for the Future* initiative designed to: take action on climate change; reform water markets to increase efficiency and productivity of on-farm irrigation; secure future water supplies via desalinization, water recycling and stormwater harvesting; and to 'buy-back' water entitlements from farmers – thus allowing water to be returned to the environment (Department of Sustainability, Environment, Water, Population and Communities, 2010b).

A key feature of the water 'buy-back' was the allocation of $3.1 billion to purchase the water entitlements from willing sellers. It was estimated that to ensure the long-term survival of the river system, some 3,500 gigalitres per year – equivalent to a 22 per cent cut in available water each year for irrigated agriculture – would be required. ABARE calculated that there would be a 15 per cent reduction in the gross value of irrigated agriculture if this amount of water were to be returned to the environment. Through the Murray-Darling Basin Authority (MDBA), an opportunity has recently been provided to interested parties to comment on the plans to divert this amount of water to the environment. The National Farmers Federation (NFF) – the peak national body of farmers – has rejected the proposed Basin plan out of hand. It argues that there would be a contraction in farming, a reduction in farm numbers, and furthermore that removing water from agriculture would have indirect, negative impacts upon country towns. It questions the science behind the MDBA's figures. It believes that the states, not the commonwealth, should be the ones determining water allocation, and it questions whether current food production levels could continue if such a large volume of water were to be removed from irrigation farming (NFF, 2010). At the beginning of 2011, the plan remained highly contentious and action on it unresolved.

While it is true that there is disagreement about the extent of the water that must be returned to the Basin's rivers to allow for environmental recovery (Rochford, 2009), it is clear that the NFF – along with farmers themselves – are, not surprisingly, livid that their future livelihoods are being 'attacked' in the name of environmental security. The federal government established the MDBA to stop vested interests from hijacking attempts to provide an overall, non-political Basin plan (Marshall and Stafford Smith, 2010); but this has not prevented a fight over the rights of farmers to the water they have previously been allocated. It is somewhat contradictory that the public debate in the region separates the natural environment from human society – with a common call that it is wrong to 'put the environment before people'. There is a notable lack of recognition that without a sustainable natural environment, human habitation and economic activities in the Basin will ultimately be compromised

## Adaptation and mitigation options under neoliberalism

Neoliberal approaches to climate change adaptation and mitigation options stress the desirability of reducing the state's role in funding and/or interfering in on-farm decision-making, and of endorsing 'self-help' solutions for individuals and communities (Lockie et al., 2005; Marshall and Stafford Smith, 2010). It is based upon an assumption that the market is the best mechanism to allocate scarce resources and that governance arrangements must create opportunities to foster market-based solutions to society's problems. In terms of mitigation (pursuing actions to reduce greenhouse gas emissions), bans on land clearing have already helped by reducing carbon dioxide emission levels by 40 per cent of 1990 levels (notably, from tough government regulation rather than market-based decision-making). But Australia does not have an emissions trading scheme and developing a price on carbon was, at the time of writing, in the political 'too hard' basket. Yet, it has been suggested that Australian farmers will have no incentive to pursue mitigation options such as leaving trees in the ground – and planting more trees – until there are funding incentives to do so. Some have questioned the ability of the market to send the right signals (see Catchpole, 2007). For example, if irrigated farmers turned pasture into forest (gaining carbon credits) there would be a subsequent decline in flow downstream (thereby lowering the farmer's water trading capacity). How would these two contradictory outcomes be reconciled to alter current farmer behaviour?

With payments for future mitigation options 'on hold', the need for adaptation within Australian farming is viewed as imperative. A number of adaptation options have been suggested, including:

Cropping and horticulture:

- plant species that have increased resistance to heat, frosts or drought
- alter application times for fertilization and irrigated water applications
- alter timing and location of cropping activities
- use varieties that are more pest and disease resistant
- retain stubble to reduce soil erosion.

Livestock:

- adapt annual production cycle to better match the production of feed
- alter pasture rotations and modify grazing times
- alter animal species or breeds
- provide supplementary feeding.

Sector-wide:

- diversify farm income and increase off-farm income
- invest outside agriculture
- offset increased cost of managing climate change by reducing other costs
- employ new risk management tools – such as futures contracts and water trading

Stokes and Howden (2008)

Most of these 'new' adaptation options are strategies that Australian farmers have been doing for decades. The most significant change is the shifting of risk from the state to individual farmers, through tools such as futures trading. Previously, such risks were shared by the state via single-desk systems such as wheat and wool boards. Under neoliberalism, however, the philosophy of self-help prevails. Placing more risk in the hands of farmers does little to respond to environmental problems associated with productivist agriculture.

Technocratic responses to climate change are the same options that have resulted in the further intensification of farming. They conform to – rather than challenge – the existing system of productivism. It should also be remembered that any fundamental alteration to existing production regimes can be difficult to achieve when farmers are in debt – as most of Australia's primary producers are.

The banks can dictate how loaned funds can be spent – and this is not, usually, for radical alterations to the productivist format – leaving the intensification of current production as a favoured bank requirement (Lawrence *et al.*, 2007).

In the case study region of the MDB, it has been predicted that 'adaptation' to climate change will include the movement away from irrigated pastures, rice production and irrigated wheat towards higher-return products of fruits, vegetables and cotton. These latter crops will require increases in the application of fertilizers and pesticides. Some areas of wheat production will change to pastoralism, but this will mean soils will be exposed to hard-hoofed animals that will potentially increase soil degradation.

In the MDB, some irrigators have responded to climate change by seeking to purchase new water entitlements, while others have sought to improve efficiency of on-farm water use by investing in pipes, centre pivots and so forth. However, the water market is not yet on a firm footing: some farmers remain uncertain of the benefits they will gain from additional water; others question whether water access in their district will be 'rationalized', with the whole district being disconnected from the irrigation water supply (Rochford, 2009). Basically, if many farmers sell their licences, this makes the supply system more expensive and therefore potentially unviable for others who want to continue using irrigation water. Farmers are confused by the signals from government and are unclear about the best course for individual on-farm adaptation. Rochford (2009) makes the perceptive, but cynical, point that, in such circumstances, most farmers are currently adapting to *policy* change, rather than to *climate* change. As he implies, the two might not be compatible.

In relation to wider governance structures, the Australian government is pursuing a devolved neoliberal approach to community actions on climate change – via collaborative partnerships and hybrid governance structures (participatory NRM Basin associations) – while centralizing control of funding (Lawrence, 2005; Marshall and Stafford Smith, 2010). It has been pointed out, however, that the two elements here are in tension. On the one hand, the government wants the NRM (Natural Resource Management) bodies to be adaptive systems that can deal with 'abrupt, turbulent and surprising' change (Marshall and Stafford Smith, 2010, p270); on the other hand they are hindering adaptation via strongly mechanistic and top-down approaches to funding and administrative support. What becomes entrenched is a system which favours incremental adaptation when what is needed is *transformation* (Marshall and Stafford Smith, 2010, p274). In theory, a neoliberalist approach to NRM *should* generate locally-based and locally-owned (devolved) approaches which have the capacity to differ across regions according to local factors and different stakeholder needs. But the historical dominance of the state in funding NRM activities, in concert with the poor funding received by, and opportunities available to, non-government sectors involved in NRM, militates against the emergence of a more flexible, responsive and collaborative-stakeholder approach at the local level (Marshall and Stafford Smith, 2010).

So should we 'kill' any state activity in NRM in the hope that new private sector/community activities will arise, phoenix-like, from current locally-based NRM? No. In Australia, the Landcare movement and regional catchment management have shown us many good things. Two critical failings, however, have been parliamentary short-term thinking and political ineptitude. Basically, there is no consensus about the likely future effects of climate change in Australia (indeed, there are still many climate change deniers in the ranks of the coalition parties), and there has been a failure by the current Federal Labour Government to give a clear direction for climate change policy. Despite the appearance that federal and state governments are joining forces in addressing NRM and climate change, the arrangements thus far have not proven to be robust and flexible enough to consistently meet the needs of catchment communities: rather, they are designed to ensure that state interests are prioritized. It could be argued that unless there is constant, long-term and committed bipartisan agreement on what is needed to support farmers' adaptation to climate change, the neoliberal agenda will predominate: farmers will be required to expand output or face the consequence of leaving the land as victims of climate change and unpredictable markets.

Finally, there is also a poor alignment between adaptation strategies at the government, regional and individual levels (Rochford, 2009). The capacity of local/regional NRM groups to move quickly to the challenges of climate change will be impeded by the structural constraints of existing forms/priorities/ideologies of government. The outcome – at this stage, at least – is conservative, incremental change that reinforces the framework of productivism, rather than looking beyond it.

Australian agriculture is faced with a choice in relation to climate change adaptation. It can adapt *within* the present system of productivism, or it can seek to transform farming by placing agricultural production on a less environmentally-destructive pathway. This is adaptation *outside* productivism. Adapting *within* productivism means leaving all the current fundamentals in place (Gray *et al.*, 2009). Farmers will use the latest high-tech products from agribusiness in a quest to achieve productivity increases, and agribusiness will facilitate this via new genetically-modified plants and animals. In the case of plants, the new generation seeds the farmers purchase will be the intellectual property of companies like Dow, DuPont and Monsanto, and they will be designed to produce the best outcomes only when expensive proprietary agrochemicals are used in tandem. The system will remain dependent upon the petrochemical industry, so the costs of farming will accelerate. Financial institutions will come to play a more important role in farming, seeking ever-higher returns for shareholders from investments in products such as beef, coffee, sugar and grains. The financialization of Australian agriculture will have a speculative element to it – leading to the increasing price of farmlands, and increases in the cost of food (Burch and Lawrence, 2009). This will do little to address issues of food security and food equity, particularly in a world where cropping and pasture land continues to decline in availability (Hanjra and Qureshi, 2010; Lawrence *et al.*, 2010). As indicated earlier, the trajectory in

Australia will be one of fewer farmers producing larger amounts of output while continuing to exploit the environment.

Adaptation *outside* productivism is something of an unknown, and will be a challenge for farmers who embark on this course. We do, though, understand some of its likely contours: decreasing reliance on synthetic fertilizers and pesticides; less dependence on commercial agribusiness firms; a move away from monocultures and the factory-farming of animals; rejection of genetically modified organisms; the application of low-input and organic farming practices; recycling; the reduction of 'food miles' via a grass-roots led re-localization of food production; the protection of local and global agro-biodiversity; and the utilization of urban space for agriculture. Many of these ideas accord with the principles of 'agro-ecology' – defined as the 'application of ecological concepts and principles to the design and management of sustainable agro-ecosystems' (Altieri, 2010). There is evidence that organic and agro-ecological methods have out-performed productivist approaches by providing environmental benefits such as soil water retention (and through this increased drought tolerance) and improvement in soil fertility (Environmental News Service, 2009). The approach has the potential to be particularly beneficial in terms of climate change as it would improve absorption and storage of carbon in plants, reduce carbon emissions from crop and livestock systems, and reduce the nitrous oxide emissions associated with the application of inorganic fertilizers (Schahczenski and Hill, 2009). It is imperative that Australian agriculture embraces approaches that use high carbon-cropping; move away from intensive livestock production; conserve the carbon currently stored in grasslands and forests; and plant new vegetation in degraded farming landscapes (Scherr and Sthapit, 2009).

The problem is that agro-ecology threatens the hegemonic position of productivist farming and appears to have little support from Australia's overwhelmingly conservative farming organizations. Governments, too, while mouthing statements about the importance of 'sustainability', 'biodiversity protection' and so forth, put very little funding into 'alternative' agro-ecological approaches to farming. Critics also say that alternative, lower impact forms of farming will not produce the volume of food to enable Australia to play its part in 'feeding the world' – particularly with a predicted growth in world population from the present 6.8 billion to 9.2 billion by 2050. For the foreseeable future, it would appear climate change adaptation in Australia will reinforce existing trends towards high-output – but unsustainable – productivist agriculture.

## Conclusion

In this chapter, we have argued that neoliberal ideology – embracing principles of individualization and marketization – has helped to generate and reinforce a particular form of farming system in Australia. It is a system based upon: product specialization; operational intensification; returns to scale (and therefore continual pressure for the expansion in size of operations); and economic concentration, with fewer farmers producing larger volumes of output. It is a system of *productivism*, underpinned by a rationalizing and legitimating ideology.

Productivist agriculture has provided domestic consumers with abundant and relatively cheap food, and has enabled the nation to export large volumes of food to overseas consumers. To the extent that it has simultaneously achieved both these outcomes, it has been viewed as a highly successful approach to farming. But its 'success' has come at a very high price to the environment, with literally billions of dollars each year spent addressing the problems associated with soil depletion, salinization, water pollution and a host of other ecological problems. We question whether a 'resilient' agricultural system can be one which destroys the environment upon which it depends.

Productivism has been premised upon continued availability of irrigated water. However, climate change scenarios for Australia point to a deteriorating and/or less reliable supply of water for many of the nation's most productive agricultural regions. Governments have, however, recognized the importance of guaranteeing long-term environmental flows to save ecosystems that have been placed under severe pressure through over-allocation of water resources to agriculture. Neoliberal approaches to water management are expected to see the creation of water markets to enable water to be used more efficiently and for the highest value farming activities. Yet the creation and expansion of markets for previously taken-for-granted natural assets like water may fail to create a more effective system for dealing with problems of water overuse and poor allocation. After all, there is a growing body of evidence to suggest that past decades of 'free market' pressures on farmers to produce more for less has been the cause of, rather than a solution to, the problem of environmental degradation.

The MDB example presented here is only one of many degraded agricultural environments in Australia. As noted earlier, Australian agriculture feeds the national population plus an additional 40 million mouths. A reduction of the amount of viable, environmentally sustainable agricultural land both in Australia and globally – coupled with the challenges of climate change and peak oil – are placing pressures on the global food supply that cannot be addressed through the productivist paradigm.

Farmers are expected to 'adapt' to climate change through a variety of incremental steps – such as altering production times and using more productive (and more pest-resistant) plants and animals. In this chapter, we argue that such incremental change will occur *within* the current productivist regime and will therefore do very little to challenge the unsustainable system that has become entrenched over the last fifty years. We contend that new options for climate change mitigation and adaptation should be considered by both farmers and governments. These options would be *outside* the system of productivism and would foster a move towards agro-ecology. Without being unduly pessimistic, we consider that, without strong government and industry acceptance of the desirability of moving to agro-ecology, there is little hope that Australia will break the hegemony of productivism, and will therefore be unlikely to escape the straightjacket that promotes higher volumes of production at the continuing expense of the environment. Ultimately, a degrading environment, poor terms of trade for farmers and the twin challenges of peak oil and climate change will have consequences for food security not only within Australia but also beyond its shores.

## References

ABARE. (2007) 'Climate change: Impact on Australian agriculture', *Australian Commodities*, vol. 14, no. 4, pp657–676

Altieri, M. (2010) 'Scaling up agroecological approaches for food sovereignty in Latin America', in H. Wittman, A. Desmarais and N. Wiebe (eds) *Food Security: Reconnecting Food, Nature and Community*, Fernwood Publishing, Halifax

Argent, N. (2002) 'From Pillar to Post? In Search of the Post-productivist Countryside in Australia', *Australian Geographer,* vol. 33, no. 1, pp97–114

Australian Government (2010) 'The need for action', www.climatechange.gov.au/publications/cprs/white-paper/~/media/publications/white-paper/V1002Chapter-pdf.ashx, accessed 5 January 2011

Beale, B. and Fray, P. (1990) *The Vanishing Continent,* Hodder and Stoughton, Sydney

Burch, D. and Lawrence, G. (2009) 'Towards a third food regime: Behind the transformation', *Agriculture and Human Values*, vol. 26, pp267–279

Catchpole, H. (2007) 'Curbing climate change', www.abc.net.au/science/features/curbing-climatechange, accessed 6 January 2011

Council of Australian Governments (1994) 'Attachment A: water resource policy', www.coag.gov.au/coag_meeting_outcomes/1994-02-25/docs/attachment_a.cfm, accessed 2 January 2011

Crabb, P. (1988) *Managing Water and Land Use in Inter-State River Basins*, Research Paper No. 1, School of Earth Sciences, Macquarie University, Sydney

CSIRO. (2008) *Water Availability in the Murray-Darling Basin: Summary of a Report from CSIRO to the Australian Government*, CSIRO, Canberra

Department of Sustainability, Environment, Water, Population and Communities (2010a) 'Water for the Future', www.environment.gov.au/water/locations/murray-darling-basin/index.html, accessed 31 December 2010

Department of Sustainability, Environment, Water, Population and Communities (2010b) 'Water for the Future: Fact Sheet', www.environment.gov.au/water/publications/action/water-for-the-future.html, accessed 3 January 2011

Environmental News Service (2009) 'The environmental food crisis: A crisis of waste', Environmental News Service, www.ens-newswire.com/ensfeb2009/2009-02-17.01.asp, accessed 20 December 2010

Gray, I. and Lawrence, G. (2001) *A Future for Regional Australia: Escaping Global Misfortune*, Cambridge University Press, Cambridge

Gray, I., Lawrence, G. and Sinclair, P. (2009) 'The sociology of climate change for regional Australia: Considering farmer capacity for change', in J. Martin, M. Rogers and C. Winter (eds) *Climate Change in Regional Australia: Social Learning and Adaptation*, VURRN Press, Ballarat

Gunasekera, D., Kim, Y., Tulloh, C. and Ford, M. (2007) 'Climate change: Impacts on Australian agriculture', *Australian Commodities*, vol. 14, no. 4, pp657–676

Hanjra, M. and Qureshi, E. (2010) 'Global water crisis and future food security in an era of climate change', *Food Policy*, vol. 35, pp365–377

Keane, B. (2008) 'ABARE ignores inconvenient truths', www.crikey.com.au/2008/06/04/abare-ignores-inconvenient-truths, accessed 20 October 2010

Lawrence, G. (2005) 'Promoting sustainable development: The question of governance', in F. Buttel and P. McMichael (eds) *New Directions in the Sociology of Global Development*, Elsevier, US

Lawrence, G. and Vanclay, F. (1992) 'Agricultural production and environmental degradation in the Murray-Darling Basin', in G. Lawrence, F. Vanclay and B. Furze (eds) *Agriculture, Environment and Society: Contemporary Issues for Australia*, Macmillan, Melbourne

Lawrence, G., Lyons, K. and Wallington, T. (eds) (2010) *Food Security, Nutrition and Sustainability*, Earthscan, London

Lawrence, G., Richards, C. and Cheshire, L. (2007) 'The environmental enigma: Why do producers professing stewardship continue to practice poor natural resource management?' *Journal of Environmental Policy and Planning*, vol. 6, no. 3/4, pp251–270

Lockie, S., Lawrence, G. and Cheshire, L. (2005) 'Reconfiguring rural resource governance: the legacy of neoliberalism in Australia', in P. Cloke, T. Marsden and P. Mooney (eds) *Sage Handbook of Rural Studies*, Sage, London

Marshall, G. and Stafford Smith, M. (2010) 'Natural resources governance for the drylands of the Murray-Darling Basin', *The Rangelands Journal*, vol. 32, pp267–282

Murray-Darling Basin Authority (2010) 'Murray-Darling Basin Agreement', www.mdba.gov.au/about/governance/murray-darling-basin-agreement, accessed 2 January 2011

National Farmers' Federation (2010) *National Farmers' Federation Submission to the Guide to the Proposed Basin Plan*, NFF, Canberra

Nossal, K., Abdalla, A., Curtotti, R., Tran, Q. and Brown, A. (2008) *GM Crops in Emerging Economies: Impacts on Australian Agriculture*, ABARE, Canberra

PMSEIC (Prime Minister's Science, Engineering and Innovation Council) (2010) *Australia and Food Security in a Changing World*, Canberra, Commonwealth of Australia, www.chiefscientist.gov.au/wp-content/uploads/FoodSecurity_web.pdf, accessed 26 December 2010

Rochford, F. (2009) 'Climate change and irrigation water delivery – substantive constraints on social learning', in J. Martin, M. Rogers and C. Winter (eds) *Climate Change in Regional Australia: Social Learning and Adaptation*, VURRN Press, Ballarat

Schahczenski, J. and Hill, H. (2009) 'Agriculture, climate change and carbon sequestration', National Sustainable Agriculture Information Service, www.attra.ncat.org, accessed 24 February 2009

Scherr, S. and Sthapit, S. (2009) 'Farming and land use to cool the planet', in The Worldwatch Institute, 2009 *State of the World: Into a Warming World*, www.worldwatch.org/stateoftheworld, accessed 24 February 2009

Spraggon, B. (2011) www.abc.net.au/rural/murraydarling/?site=rural&microsite=murraydarling&section=article&date=(none)&page=8

Stokes, C. and Howden, S. (eds) (2008) *An Overview of Climate Change Adaptation in Australian Primary Industries – Impacts, Options and Priorities*, CSIRO, Australia, www.csiro.au/files/files/plhg.pdf, accessed 5 January 2011

# 10

# FOOD SECURITY AND THE DE-AGRARIANIZATION OF THE INDONESIAN ECONOMY

*Jeff Neilson and Bustanul Arifin*

Food insecurity and the spectre of famine have been persistent, and sometimes volatile, political issues in Indonesia since the country gained independence in 1945. During the early 1960s, a combination of drought; a rat plague on Java; the destruction of crops due to the eruption of the Gunung Agung volcano on Bali; and imprudent economic policy resulted in large-scale food shortages across the archipelago. A 1964 article in *Time Magazine*, 'Indonesia: Of Rice and Rats', described the dire situation on Java:

> Nearly 1,000,000 people were on a starvation diet in Java; scores have already died of malnutrition. Peasant villages emptied as food supplies dwindled, and native families poured into already overcrowded cities.

This same period of chronic food insecurity was portrayed in the 1982 film, *The Year of Living Dangerously,* when Billy Kwan, the half-Chinese, half-Australian cameraman working with journalist Guy Hamilton, unfurls a protest banner from a Jakarta building reading 'Sukarno feed your people'. In the film, Billy is subsequently killed as a result of this provocative political statement. In reality, President Sukarno went on to lose his grip on power due, at least in part, to his inability to ensure food security for the population. Sukarno's successor, President Soeharto, subsequently made national food security (narrowly equated with national-scale rice self-sufficiency) a central pillar of economic and social policy. Indeed, ramifications from the 1964 rice crisis and associated food riots are still felt in the political formulation of food policy in Indonesia today.

For Soeharto, rice self-sufficiency reflected a dominant international perspective on food and hunger, widely held until as recently as the mid-1990s, which prioritized production-centric explanations aimed at maintaining domestic food stocks. The work of Amartya Sen, particularly his 1981 book *Poverty and Famines: An Essay on Entitlement and Deprivation*, however, led to a fundamental shift in the way that scholars and policymakers approached food and famine. A highly simplified rendering of Sen's work explains that famine is not a result of food shortages per se, but is due to the absence of resources (or entitlements) at the individual or household level to access food. The 1996 World Food Summit in Rome was later widely accredited with introducing a framework for understanding food security that incorporates Sen's three dimensions of availability, accessibility and utilization. As this chapter demonstrates, the discursive practices that shape food policy in Indonesia, however, have been largely unable to accommodate such a household perspective. Reconciling policy priorities at different scales (national, regional and household) by managing the relationships between supply control and household access to food lies at the heart of the food security conundrum in a de-agrarianizing Indonesia, and is the core concern of this chapter.

Soeharto ruled Indonesia autocratically for 32 years, from 1966 until 1998, in what became known as the *New Order* regime. The New Order is generally associated with the introduction of pro-western economic liberalism into Indonesia; rampant and highly centralized corruption; severe limitations on individual democratic freedoms; and the sometimes violent repression of activist movements and human rights. Peasant organizations and others advocating land reform were singled out for particularly harsh treatment by the regime due to their association (real or imagined) with the outlawed and demonized Communist Party of Indonesia (PKI). Despite this extreme political marginalization, it is inaccurate to portray the regime as systematically working against the commercial interests of farmers. Economic policies during this period even indicated a structural bias towards agriculture in an attempt to ensure political stability and stimulate food production through farmer appeasement (Timmer, 2004). The regime, however, tended to equate food security with national-level rice self-sufficiency (i.e. an emphasis on a narrow definition of food 'availability'), maintaining a single-minded policy of rice protectionism.

Structural changes in the Indonesian economy, associated with socio-economic processes of de-agrarianization, now demand new strategic approaches to food security, requiring greater emphasis on food accessibility at the household-level and an appreciation of the diverse livelihood portfolios of the Indonesian poor. The emergence of representative democracy in Indonesia since 1998 (a period generally referred to as *Reformasi*) has, instead, served to strengthen the political case for food protectionism. Indeed, subsequent administrations have responded to nationalist political demands for domestic food production by maintaining various agricultural subsidies and extending a protectionist trade policy. Food self-sufficiency, and the specific scale at which this should be achieved, continue to be highly politically contested policy goals in Indonesia; goals informed by the discursive practices of various interest groups both within and outside of the archipelagic state.

This chapter argues that, despite some initiatives to improve food accessibility over the last decade, Indonesian policy continues to conflate food security with rice self-sufficiency, often to the detriment of food-insecure households. As the evolving nature of the Indonesian economy moves the country further away from being a predominately agrarian nation, differing visions of food security are now being brought into sharp contrast. Despite insistent claims that rice protectionism actually increases poverty (such as those expressed by Warr (2005)), a radical political departure from current policies is unlikely in the foreseeable future. Complicating food security policy options further, the relative resilience of the Indonesian food system during the 2008 global food crisis (primarily due to strong domestic supply) has provided grist for the mill for advocates of enhanced food protectionism in the country.

## Food insecurity in Indonesia

In 2008, the National Food Security Agency (Dewan Ketahanan Pangan (DKP)) classified approximately 87 million (36 per cent) of Indonesia's 240 million people as food-insecure, of which 25 million were severely food-insecure (Martianto, 2009). Clearly, food security remains a daily challenge for Indonesia and many of its people.

The United Nations Millennium Development Goals include a target to halve the proportion of people suffering from hunger globally by 2015 (from 1990 levels). In Indonesia, the indicators used to measure progress towards this goal are the percentage of children under five suffering from severe malnutrition (severely underweight) and undernourishment (moderately underweight). In 1989, 37.5 per cent of Indonesian children were suffering from severe malnutrition and under-nourishment (Bappenas, 2007). While this had decreased to 26.4 per cent in 2002, it had risen again slightly to 28.17 per cent in 2005, suggesting that Indonesia is unlikely to meet its target of 18.7 per cent by 2015 (Bappenas, 2007). Neither is undernourishment solely a concern for infants in Indonesia. According to the United Nations Food and Agricultural Organization (the FAO), in its 2010 State of Food Insecurity report, 13 per cent of Indonesia's entire population was under-nourished during 2005–2007.

Food insecurity is also unevenly distributed across Indonesia. Indonesia's Food Security and Vulnerability Atlas (FSVA) jointly released in 2010 by the World Food Programme (WFP) and the National Food Security Agency, identifies 30 districts (out of the then 346 districts in total) as receiving Priority 1 ranking according to its Composite Food Security Index — and thus experiencing 'chronic food inse-curity'. Of these 30 districts, 16 were concentrated in Indonesian Papua and six were in Nusa Tenggara Timur (NTT, part of the island chain between Bali and East Timor). Poverty is the primary factor resulting in food insecurity in Indonesia, with these same provinces experiencing the highest poverty rates in Indonesia — with Papua (see Figure 10.1) the highest at 40 per cent (DKP, 2010). However, the relationship between poverty and prevalence of undernourishment is by no means entirely linear. The Special Province of Yogyakarta on Java experiences levels of poverty above the national average but low levels of undernourishment, while other

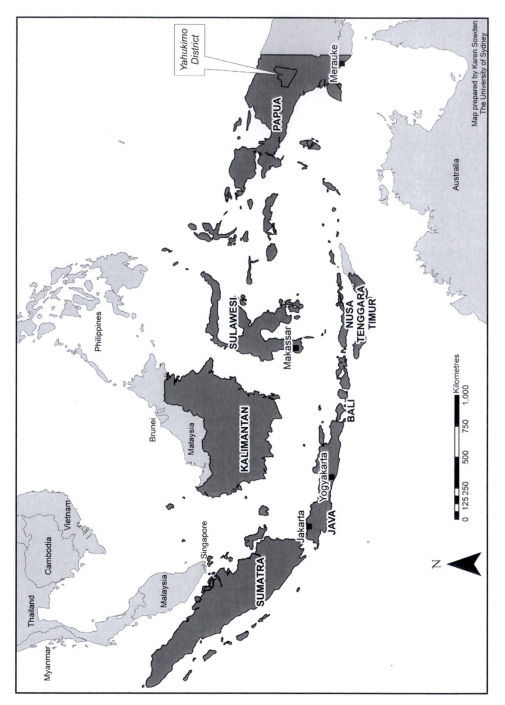

*Figure 10.1* Map of Indonesia including Yahukimo District

provinces, such as South Kalimantan, experience high levels of undernourishment amongst relatively low levels of poverty (Bappenas, 2007). Makassar, the capital of South Sulawesi province — a rice surplus province — has been widely reported as experiencing numerous fatalities due to malnutrition (Hajramurni, 2008). Such apparent anomalies demand solutions beyond a sole focus on either food self-sufficiency (at any scale) or poverty reduction alone.

While food aid for disaster relief has, tragically, become somewhat routine in Indonesia in recent years, with natural disasters such as the Indian Ocean tsunami, the Yogyakarta and Nias Island earthquakes, and the 2010 volcanic eruption of Mt Merapi, reports of famine and severe malnutrition unrelated to such events continue to cause great concern. In late 2009, for example, reports from Papua emerged of 113 deaths in Yahukimo District due to hunger and associated diseases when the local sweet potato harvest failed (AHRC, 2009). In the Yahukimo case, despite initial government denials that lack of food led to the deaths, it became apparent that failures in broader food security institutions were indeed to blame. Elsewhere in Indonesia, urban poverty and the chronic inability of the poor to afford basic foods have resulted in what some claim to be an epidemic of 'hidden hunger'. Newspaper reports from South Sulawesi suggested that over a two-month period in 2008, there were ten infant deaths from hunger and 94 other cases of acute malnutrition in the provincial capital of Makassar (Hajramurni, 2008).

Causes of severe hunger are clearly diverse and defy simple categorization or linear associations. It is therefore somewhat disturbing to observe policy formulation occurring based on assumptions that a direct relationship exists between food availability (domestic production) and food security in Indonesia. Certainly, there is a deeply-held belief amongst most Indonesian people and policy-makers that rice self-sufficiency equates to food security — a sense that was exploited by the New Order regime to gain widespread support for its quest to achieve national self-sufficiency in rice, and which continues to influence current debates on national food policy.

## At what cost should Indonesia strive to feed itself?

During the 1970s and 1980s, Indonesia achieved remarkable progress towards improving the availability of food domestically through a combination of agricultural supports, protectionist trade policy, public purchasing and the maintenance of buffer stocks. As a result of these policies, Soeharto received an award from the FAO during the 1985 World Food Summit for successfully transforming his country from being the largest rice importer in the world to being self-sufficient. This achievement was widely celebrated within Indonesia, fuelling a strong sense of nationalistic pride and instilling a deeply-held belief in the population of the inherent cultural significance of a nation being able to feed itself. Everyday Indonesians recall staged photos of President Soeharto wearing a farmer's hat with a hoe in his hand, amidst green fields of rice, heavy with grain. Domestic rice production actually fluctuated slightly above or, more commonly, below

*Table 10.1* Production of major food commodities in Indonesia, 2006–09

| 'Strategic' Crops | 2006 | 2007 | 2008 | 2009 |
|---|---|---|---|---|
| *Rice* | | | | |
| Productivity (ton/ha) | 4.62 | 4.71 | 4.89 | 4.94 |
| Production ('000 ton dry unhusked) | 54,455 | 57,157 | 60,326 | 62,561 |
| *Corn* | | | | |
| Productivity (ton/ha) | 3.47 | 3.66 | 4.08 | 4.16 |
| Production ('000 ton dried grain) | 11,609 | 13,288 | 16,317 | 17,041 |
| *Soybean* | | | | |
| Productivity (ton/ha) | 1.29 | 1.29 | 1.31 | 1.32 |
| Production ('000 ton dried grain) | 748 | 592 | 776 | 925 |
| *Sugarcane* | | | | |
| Productivity (ton/ha) | 5.90 | 6.08 | 6.20 | 6.26 |
| Production ('000 ton sugar) | 2,267 | 2,402 | 2,542 | 2,840 |

Source: Central Agency of Statistics (BPS), various issues

national consumption levels in the following years, until massive deficits were again recorded in 1998.

### Subsidizing rice production

The flagship agricultural development programme during the New Order was the *Bimbingan Massal* or Mass Guidance Scheme (BIMAS), which focused on enhancing domestic rice production. BIMAS encompassed a package of integrated farmer support structures: funding for improved rural infrastructure such as irrigation and roads; technology transfer through agricultural extension; subsidized agricultural credit channelled through village-based cooperatives; the generation and dissemination of improved 'green revolution' crop technologies; and provision of agricultural machinery, such as tractors and hand sprayers, to farmers. Critically, the government ensured the availability of low-price fertilizers by supporting the establishment of state-owned factories in strategic locations across the country. Removal of the costly fertilizer subsidy was included within the austerity measures specified within the Letter of Intent (LoI) that Indonesia signed with the International Monetary Fund in 1998 (pesticide subsidies had already been abolished as early as 1989). Popularly-elected governments subsequently re-implemented the fertilizer subsidy in 2003.

The reformasi period has delivered a new national institution, the National Food Security Board (Dewan Ketahanan Pangan (DKP)), which was established in 2001 by Presidential Decree (No. 132), with an implementing Presidential Regulation (No. 83) issued in 2006. The DKP was tasked to assist the president to formulate a national food security policy which would ostensibly cover various areas such as food availability, maintenance of food stocks, distribution, consumption, food quality and variety, nutrition, and food safety. While the inclusion of food consumption and nutrition within the DKP would appear to suggest greater attention to household-level food accessibility and utilization, in reality a core concern for national-level food self-sufficiency has been retained. This is perhaps not surprising, as the DKP is supported by a full-time secretariat within the Ministry of Agriculture. This institutional location challenges the impartiality of the DKP at moments when it needs to carefully balance consumer and producer interests instead of responding to its 'natural' farmer-based constituency.

During the 2004 presidential election campaign, Susilo Bambang Yudhoyono (SBY) highlighted his recent doctoral study in rural political economy from Bogor Agricultural University, leading to widespread expectations of enhanced agrarian support during his presidency. Indeed, agricultural production and productivity did increase during his first presidential term (Table 10.1), with claims again being made in 2008 that Indonesia was self-sufficient in rice. An important strategy to bolster food production during the period 2005–2009 (as reported in the 2010 Food Security and Vulnerability Atlas, in DKP (2010)) was the expansion of permanent agricultural land, accompanied by a two-fold increase in the budget allocation for fertilizer subsidies; enhanced regulation of LPG for fertilizer production; and agricultural investment and credit incentives (FSVA, in DKP (2010)). These initiatives reflect the central aim of the DKP, explicitly expressed as achieving (or maintaining) national self-sufficiency in 'strategic' commodities such as rice, maize, soybeans, sugar and meat.

As explained below, increasingly costly supports are required to encourage Indonesian farmers to continue growing enough rice for Indonesia to feed itself, causing some to question the actual cost-effectiveness of such a strategy.

---

### The Merauke Integrated Food and Energy Estate: food self-sufficiency at all costs?

Perhaps the most audacious and impudent policy consequence of a production-oriented approach to food security, politically invigorated by the 2008 global food crisis, has been the move to establish the Merauke Integrated Food and Energy Estate (MIFEE) in the relatively underpopulated eastern province of Papua. In a move reminiscent of the failed attempt by Suharto to establish a multi-million hectare food bowl in the peat swamps of Central Kalimantan in the 1990s (the so-called mega-rice project), MIFEE has lofty goals of generating 1.95 million tonnes of rice, 2.02 million tonnes of corn, 167,000 tonnes of soybeans, 2.5 million tonnes of sugar and 937,000 tonnes of crude palm oil each year on 1.2 million hectares of land (Deptan,

2010). The estate was officially launched in August 2010 amid hyperbolic sloganeering about the need to secure domestic food production and ensure food security for the nation. According to the Department of Agriculture, 'Merauke will not only become the national food bowl, but it will also be capable of feeding the world' (Deptan, 2010). Despite the official launch of the project, there are still considerable concerns over the practical future of the project, not least from anxious investors concerned over long-term tenure arrangements in Papua.

Not surprisingly, opposition to the Merauke plan has been overwhelming. Many NGO (non-governmental organization) observers consider the project to be nothing less than a thinly-veiled resource grab by domestic and international investors, a view not helped by the list of companies associated with the plan, including oil palm giants such as PT Medco, PT Bangun Cipta and PT Wilmar (Setneg, 2010). Serious questions have arisen regarding the environmental and carbon implications of the project, with a leading Jakarta-based environmental NGO, Greenomics, using satellite imagery, combined with data from the Ministry of Environment and the Ministry of Forestry, to estimate that no more than 500,000 hectares of suitable land (unforested scrubland classified as production forest) actually exists in the district. The implication is that MIFEE inevitably requires extensive land clearing (with the resulting release of substantial terrestrial carbon stocks). Moreover, Indonesian environmental NGOs such as WALHI predict a worsening local situation of food insecurity as a result of MIFEE as traditional resource access rights, including those allowing for sago production, are denied as a result of the project (Rakyat Merdeka, 2010). With ethnic tensions already severely strained in the province amidst heightening calls for a moratorium on further in-migration from other parts of Indonesia, it is highly ironic that the rhetoric of national food security is being invoked to justify actions likely to further marginalize traditional communities living in this critically food-insecure province.

The case of MIFEE suggests that the single-minded policy aim of ensuring food availability at the national level, and ignoring drivers of food insecurity at the household level, may have disastrous consequences.

## BULOG – The national food logistics agency for rice procurement and distribution

Complementing the network of producer subsidies during the New Order was a pricing policy implemented through the para-statal National Food Logistics Agency (*Badan Urusan Logistik,* or BULOG). A primary function of BULOG, from the time of its inception in 1967, was to maintain a floor price for rice (the

*Harga Pembelian Pemerintah* (HPP)) at the farm-gate (or close to it). The HPP was politically stipulated at a level above the estimated costs of production, with BULOG purchasing rice not otherwise absorbed by the open market in an attempt to build a national buffer stock. Initially, market intervention through the HPP was popularly interpreted as directed toward improving the welfare of Indonesian rice farmers. In the national consciousness, rice farmers were considered to constitute the marginalized backbone of a fundamentally agrarian nation and therefore deserving of support. Indeed, ensuring that agricultural produce (but not only rice) remained competitive in local and international markets was integral to Indonesia's relative success in achieving pro-poor economic growth during the New Order (Timmer, 2004). Significant government resources were allocated to supporting the notion that 'rice was the barometer of the economy', and therefore justified special policy treatment.

In addition, BULOG would intervene to release rice onto the market when consumer prices moved above a predetermined ceiling, purchasing rice on the world market when required. For many years, BULOG maintained a monopoly on rice imports, until this was dismantled as part of the 1998 IMF-imposed austerity measures. This ceiling price (enforced through *operasi pasar* or 'market operations') ensured rice was affordable to low-income households, especially in urban areas. The argument in support of such market operations was to avoid price spikes, which could trigger social unrest similar to that experienced by the previous regime. However, as Indonesian economic policy shifted towards a greater degree of openness in the early 1990s, the centralized, bureaucratic and monopolistic management system within BULOG attracted civil society criticism amidst allegations of inefficiency and corruption. In this context, price stabilization policy was becoming increasingly expensive and difficult to justify. BULOG was subsequently transformed into a state-owned enterprise in 2003, and expected to function as a profit maximizing body engaged in trading activities and warehousing, together with other commercial activities. The boundary, however, between BULOG's commercial mandate and its traditional social responsibility role remains wafer thin. Its role is also frequently confused in public debate where it is still often associated with ensuring price stabilization.

BULOG continues to be the key institution providing government rice supplies for military and civil service rations. It has also assumed responsibility for government programmes such as the RASKIN rice-for-the-poor programme (discussed below), a programme that consumed a full 80 per cent of BULOG's rice stocks in 2007 (SMERU, 2008). However, by 2008, BULOG was responsible for only procuring approximately 8 per cent of the country's rice harvest, which it stored in its 1,575 warehouses with a total storage capacity of nearly four million tons — compared with national production in 2008 of 37 million tons (Arifin, 2009). While the overall cost of BULOG's role as a national logistical warehousing agency, relative to its social benefits, continue to be sharply debated in Indonesia, it is generally accepted that the agency was successful in its role of stabilizing rice prices during the New Order regime (as argued by Timmer, 1996).

## *Trade policy and food security*

Throughout the New Order, BULOG was granted a monopoly over imports of all major tradeable foodstuffs — rice, sugar, maize and soybeans. And for much of this period, domestic rice prices were actually maintained in relative accordance with world prices (Thomas and Orden, 2004). The large national rice deficits recorded in 1998, combined with the IMF structural adjustment package, even resulted in the temporary liberalization of rice imports. Starting in 2000, however, the popularly elected governments of Presidents Abdurrahman Wahid and Megawati Sukarnoputri began to again tightly regulate imports to protect domestic rice producers. This was achieved through a formal import tariff of 25–30 per cent. (Thomas and Orden, 2004), however, estimate that effective tariffs were as high as 75 per cent at that time due to non-tariff barriers such as difficult customs regulations and costly inspections.) Special licensing regulations also limited the involvement of private sector traders, reducing competition and possibly resulting in further upward pressure on prices. This policy was then followed by a temporary ban on all rice imports during the Megawati presidency, before returning to a reduced import tariff (of between 8 per cent and 11 per cent) during the SBY presidency.

Fane and Warr (2008) explain the tendency for protectionist trade policy during the reformasi period as resulting from the greater political influence of farmers in parliamentary democracy at a time when the ascendancy of economic nationalism in parliament corresponded with the declining influence of economic technocrats. As a result of these protectionist trade policies, and in sharp contrast to the situation during the New Order, domestic prices began to significantly diverge from world markets — the implication being that Indonesian consumers were paying the price for supporting domestic rice farmers. Indonesia is an active member of the G33 of developing countries, and is currently leading negotiations in the World Trade Organization (WTO) for the inclusion of certain special products (SPs) to be exempt from tariff reduction commitments. To this end, Indonesia is arguing for the right to independently regulate production and trade in commodities such as rice, corn, soybean and sugar cane, due to their 'special significance' for food and livelihood security. As highlighted by Fane and Warr (2009), these four commodities are also import-competing products in Indonesia.

The costs to implement the food security strategies of the New Order (especially producer subsidies and BULOG's pricing policy) were extremely high and were possible essentially through windfall government revenues generated by oil exports. Depletion of known oil reserves, stagnating exploration and rapidly rising domestic consumption have all resulted in Indonesia (a founding member of OPEC (Organization of Petroleum Exporting Countries)) being a net oil importer since 2004. The implications for national fiscal policy have been profound. With increasingly scarce budgetary resources, pressure has been brought to bear on dismantling the three core state subsidies retained in post-reformasi Indonesia — fuel, electricity and fertilizers. The changing structure of the Indonesian economy has further brought into question the immediate social benefits, relative to cost, of a strategy of national self-sufficiency in basic foodstuffs.

## De-agrarianization and the contested rural futures of Indonesia

Rural livelihoods in Indonesia, and poverty alleviation pathways more broadly, are becoming increasingly delinked from agriculture, as off-farm income generation and rural-urban migration assume ever greater significance. While this does not negate the disproportionately large number of Indonesian poor still dependent on agriculture for their livelihoods, or the importance of agriculture in facilitating pro-poor growth, it does pose a challenging set of issues for the Indonesian government when trying to formulate effective food security policy. It is, for instance, questionable whether a pro-poor vision of development linked to agriculture is desirable in the context of contemporary Indonesia. Furthermore, a national goal of rice self-sufficiency has been criticized as no longer having 'a serious macroeconomic justification' in Indonesia (McCulloch and Timmer, 2008). This goal is seen to increasingly conflict, and competes for resources, with other government goals, such as poverty alleviation, decentralization and diversification of the agricultural sector toward higher value crops.

As the points above indicate, the structure of the Indonesian economy is undergoing a process that may be tentatively referred to as de-agrarianization. According to the national employment survey (SAKERNAS), agriculture (broadly defined to include aligned sectors such as forestry and fisheries) constituted the primary source of income for 41 per cent of all households in Indonesia during 2009 and only 15 per cent of GDP (BPS, 2010). The changing structure of the Indonesian economy away from a dependence on agriculture, and rice in particular, has challenged the pro-poor assumptions of rice protectionism. It is estimated that only seven per cent of the total population of Indonesia is engaged in rice production (Warr, 2005). Even though most of Indonesia's poor live in rural areas and work in agriculture, two-thirds of farmers are actually net consumers of rice (McCulloch, 2008). Reflecting this trend, the share of agricultural value-added from rice production has shrunk from 46 per cent in 1971 to 31 per cent in 2000, while higher-value fruit and vegetables increased from 14 per cent to 22 per cent and livestock from 0.6 per cent to 5 per cent (Fane and Warr, 2009). According to these studies, policies (such as import restrictions) aimed at artificially raising domestic rice prices to encourage production and ostensibly address national food security are actually counterproductive in terms of poverty alleviation and food security at the household level. As the opening quote from the 1964 *Time Magazine* article suggests, Indonesians have, for many years now, sought solutions to food insecurity problems (at the household level anyway) in urban centres, or at least through diversified rural incomes rather than through household-level self-sufficiency.

Do these structural changes in the Indonesian economy require a new strategy for food security? Throughout the 1980s and 1990s, ensuring that the resources held by the poor (their own labour and sometimes their own land) could be converted to profitable economic activities through prudent macroeconomic policy was a key strategy to ensure pro-poor economic growth in Indonesia (Timmer, 2004). Agriculture was obviously an important part of this equation. But subsidized fertilizers, primarily intended for rice farmers, were easily applied to other, more lucrative crops. Over time, poverty alleviation in rural areas was increasingly achieved through export-oriented agricultural crops, such as coffee and palm oil in Sumatra, cocoa in Sulawesi, and high-value horticultural crops in Java, and then again by the growth of off-farm rural incomes. The impressive achievement of cocoa-derived poverty alleviation (and enhanced food security) in rural Sulawesi during the 1980s and 1990s is a case-in-point (Ruf and Yoddang, 2001). In a recent study by the authors of livelihood strategies for coffee farmers in Sulawesi and Nusa Tenggara (Neilson *et al.*, 2011), food insecurity was found to be much higher in Nusa Tenggara, where, paradoxically, food production was a far more central component of livelihoods. In this study, the most food-secure districts were those in Sulawesi, where farmers were engaged in a variety of commercially-oriented agricultural pursuits (as in Enrekang District) and where 'farmers' were highly dependent on off-farm income, especially remittances (as in North Toraja District). If, at the household scale, there is little correlation between engagement in primary food production and food security, is there any rationale at the national scale for supporting policies that promote self-sufficiency?

### *Food sovereignty and agrarian activism in Indonesia: an alternative perspective*

Popular support for production-oriented solutions to food security has been bolstered by the resurgence of farmer organizations as a political force in Indonesia following a thirty-year hiatus. Throughout the New Order, the activities of genuine peasant movements in Indonesia were violently curtailed in the after-math of the communist purges of 1965–66, when grass-roots organizations such as the *Barisan Tani Indonesia* (Indonesian Peasants' Front) were accused of having communist allegiances. Instead, all Indonesian farmers were channelled into the *Himpunan Kerukunan Tani Indonesia* (HKTI, Indonesian Peasants' Harmony Association) — frequently managed by military figures and strongly associated with the ruling GOLKAR party. During the *reformasi* era, a number of resurgent peasant movements (*serikat petani*) have gained significant political influence, foremost among which is the *Federasi Serikat Petani Indonesia* (FSPI, Federation of Indonesian Peasants' Union, and now known simply as SPI) along with its various provincial chapters. SPI has also developed strong global networks by being aligned with the transnational peasants' movement, La Via Campesina (with SPI leader, Henry Saragih, the current Secretary General).

SPI has adopted a policy stance that critiques the FAO vision of food security, as expressed in the 1996 World Food Summit, as failing due to its reliance on biotechnological solutions and international free trade (SPI, 2008). Instead,

SPI promotes the concept of 'food sovereignty', as articulated internationally by La Via Campesina, for Indonesia. According to SPI (2008), food sovereignty for Indonesia involves protecting domestic food production through a trade policy that rejects free trade agreements on agricultural products and insists on: removing agriculture from the WTO; enforcing domestic market controls; maintaining production subsidies; localizing food self-sufficiency; introducing disincentives for export-oriented agriculture; and implementing widespread land reform. SPI successfully taps into the deeply-held popular support for rice self-sufficiency, thus ensuring political support for food protectionism.

At the heart of the agrarian future envisioned by SPI is the need for political action on land reform, a process that commenced in 1960 with the passing of the (still current) left-leaning Basic Agrarian Law, but was subsequently stymied under the New Order. For the first time in nearly 40 years, substantial pledges for agrarian reform were announced by President Yudhoyono during a speech on January 31, 2007, including the redistribution of approximately 9.25 million hectares of land (8.15 million hectares in Java and 1.1 million hectares off-Java) to landless peasants and *petani gurem* ('impoverished peasants', usually defined as cultivating less than a 0.5 hectare threshold). Yudhoyono outlined how these lands would predominantly be former forestry lands or unused plantation leases. These announcements respond directly to concerns over the poor viability of current smallholdings in Indonesia.

There are about 24.9 million agricultural households in Indonesia today (40 per cent of the total number of households), 13.6 million of which are in Java (Table 10.2). Despite a declining overall proportion of households engaged in agriculture, the absolute number has increased since 1993 (Table 10.2), with an average of around 400,000 new farm households per year. The expansion of agricultural land has occurred less rapidly than the growth in the number of farm households, such

*Table 10.2* Agricultural households in Indonesia, 1993 and 2003

| Items | Number of Agricultural Households (000) | % of farm households with <0.5 ha of land |
|---|---|---|
| *1993 Census* | | |
| Java | 11,671 | 69% |
| Off Java | 9,116 | 30% |
| Indonesia | 20,787 | 52% |
| *2003 Census* | | |
| Java | 13,583 | 72% |
| Off Java | 11,286 | 30% |
| Indonesia | 24,869 | 53% |

Source: Agricultural census of BPS, 1993 and 2003

that land fragmentation has increased (in Java, households with agricultural land of less than 0.5 ha increased between 1993 and 2003, see Table 10. 2). According to the Agricultural Census, the majority of farm households were managing agricultural land smaller than 0.5 ha in 2003 (*petani gurem*). These small land holdings ensure that it is difficult for farmers to attain economies of scale, unless the farming activities are functionally consolidated (see Siregar, 2007).

Land reform, of course, speaks to a rural future where wealth creation (or at least poverty alleviation) is fundamentally linked to access to this most basic resource. The significance of land to alleviating poverty, however, is contested by research highlighting the growth of off-farm rural incomes and rural-urban migration (Rigg, 2006). Even in the outer islands of Indonesia, where land is relatively abundant, access to land is sought primarily to grow commercially-oriented crops such as coffee, cocoa, palm oil and pepper. In contrast, the SPI vision of food sovereignty stresses the need for land ownership, localized food self-sufficiency and movement away from export-oriented agriculture.

## Scalar questions of food security in Indonesia

Indonesia underwent a massive political shift towards decentralization following the passing of Regional Autonomy Laws in 1999. The implementation of regionalized food security policy in Indonesia that resulted from these laws brings a number of common assumptions about food self-sufficiency and food security into sharp focus. If food self-sufficiency is a broadly-accepted policy goal, as it continues to be in Indonesia, what is the appropriate scale at which a geographical region should be self-sufficient in a tradeable commodity such as rice or corn? SPI argues for local self-sufficiency, while the DKP's accounting is done at a national scale. It can be argued that by even concentrating on the widely-accepted national scale, we would already be restricting the potential efficiency gains from international trade. However, with ever-decreasing scale, there are fewer real links between self-sufficiency and food security, until arguments apparently become nonsensical at the household level in the modern economy, where most food- secure urban households are unlikely to seriously consider producing their own food.

The fundamental realignment of the economic and political relationships between Jakarta and the regions, triggered by regional autonomy, helped flavour the direction of a 2002 Government Regulation on Food Security (prepared as the implementing regulation for an earlier 1996 National Food Law). This government regulation provides the framework within which local governments are expected to contribute to national food security objectives through the establishment of district-based Food Security Councils. The effectiveness of District Food Security Councils under regional autonomy has been highly varied, with most districts tending to engage in ineffectual, and costly, food staple procurement and storage activities. Another example of the localization of food security efforts in recent years has been a flagship national programme of the Ministry of Agriculture, the *Desa Mandiri Pangan (*or Self-sufficient Food Villages) through which individual villages are encouraged to strive for food self-sufficiency.

Decentralization of food security strategies through these policies has helped address the long-standing policy bias towards rice at the expense of other food crops. For many communities in eastern Indonesia, rice has never even been a staple food crop. For centuries, the sago palm has been dominant in parts of Sulawesi, the Maluku islands and coastal Papua; sweet potato, taro and yams have been integrated within highland agricultural systems of Papua possibly for millennia; and corn has been central to food security in parts of Nusa Tenggara since it was introduced from the New World. Some of these regions perceived a kind of 'rice imperialism' during the New Order regime, when consumption of these alternative foods was explicitly linked with primitivism and underdevelopment — despite being apparently well-adapted ecologically. This 'imperialism' manifested itself through government supports targeting rice production only, as well as through the longstanding, but controversial, transmigration policy of settling Javanese and Balinese rice farmers in the outer islands with the explicit aim of introducing 'superior' agricultural techniques. It also manifested itself culturally through the attitudes of government officials and mainstream media's reification of rice as a symbol of cultural sophistication. Since reformasi, however, the central government has recognized the benefits of encouraging production of alternatives to rice in these regions. Staple food diversification has even been formally promoted through a Presidential Regulation (No. 22 of 2009) aimed at accelerating the diversification of food consumption based on local resource availability.

### *Addressing food accessibility for households in Indonesia: the RASKIN rice-for-the-poor programme*

The preceding discussion has emphasized the continued policy preference for a production-oriented approach to food security in Indonesia (stressing food availability at various scales). This preference is encapsulated both in a rural future of localized food self-sufficiency, as envisioned by SPI, as well as in the grandiose food estates of Central Kalimantan in the 1990s, and MIFEE more recently in Papua. This is not to say that food accessibility has been entirely neglected in the Indonesian policy arena. During the New Order, BULOG played a key role in stabilizing food prices at the consumer level to promote accessibility. Following the 1997/98 financial crisis, and more critically following the declining fiscal importance of oil revenues, Indonesia has been required to adopt policies and programmes that more strategically target intended recipients. Politically, social welfare programmes have been necessary to offset the poverty impacts resulting from the gradual dismantling of fuel subsidies. Foremost amongst these are the Direct Cash Transfers (*Bantuan Langsung Tunai*); the PNPM Mandiri Community Development Programme (*Program Nasional Pengembangan Masyarakat*); and the Rice-for-the-Poor programme (*Beras Miskin*, or *RASKIN*), along with free education and the improvement of healthcare facilities for the poor, all of which have been put in place to explicitly support an improvement in the state of food security in the country.

The *RASKIN* programme, introduced in 2002 as a substitute for BULOG's 'market operation' ceiling price policy, is designed to reduce the severe impacts of economic crisis by providing a ration of medium-grade rice (varying, but set at about 20 kilograms per month) to targeted poor households. *RASKIN* is jointly implemented by BULOG and managed by Indonesia's National Development Planning Board (Bappenas). In 2007, the programme provided 1.9 tonnes of rice for 15.8 million poor households at a total cost of US$690 million to the national budget, with further additional costs incurred by regional governments to implement distribution (SMERU, 2008). Beneficiary households were initially identified by the National Family Planning Agency (BKKBN) based on indicators such as food intake, housing, clothing, and medical needs, and religious practices. More recently, results of the BPS Household Socio-economic Survey (SUSENAS) have been used. Initially introduced as an emergency-relief measure, it has since become a core social welfare activity, offering an alternative food security solution to rice price stabilization at the household level.

The shift in focus from nationwide input subsidies and BULOG consumer price regulation towards targeted household programmes such as RASKIN is stark. However, implementation of such targeted programmes carries its own administrative burdens. In a report prepared at the request of Bappenas, SMERU (2008) concludes that the RASKIN programme has 'relatively low effectiveness' due to inefficiencies associated with allocation of rice from the primary distribution point to the beneficiaries, inaccurate targeting of poor households and ineffectual monitoring. The World Bank estimates that only 18 per cent of the total programme budget actually reaches poor households, whereas 52 per cent benefits non-poor households, and the remainder is consumed by operational expenses and to ensure the profitability of BULOG, which is now managed as a state-owned enterprise (World Bank, 2005). The period since 1998 has therefore involved a tentative policy shift towards assisting household-level food accessibility through social welfare provision, while maintaining a strong national focus on maintaining production of strategic food crops.

## Conclusions: lessons from the 2008 food crisis for building the resilience of national food systems

The preceding discussion has emphasized the relatively high financial and social cost of pursuing food self-sufficiency goals at the national scale, particularly in a country experiencing a strong agrarian transition towards high-value horticulture and off-farm incomes. It has stressed the importance of addressing food insecurity as a phenomenon that ultimately affects households rather than nations. It also explains recent food policy trajectories in Indonesia as the product of past collective national experiences of food crises, deeply-ingrained cultural values and the rising political influence of farmers' organizations. Ultimately, however, the costs of maintaining this policy approach need to be balanced against the insurance it provides to withstand external perturbations in the world food system.

The future uncertainties that surround global food supply under changing climatic conditions, combined with the Malthusian challenges of feeding a global population set to pass seven billion in 2011, indicate a higher level of risk than ever before. Under this scenario, the critical challenge becomes how to incorporate risk and build resilience in localized and national food security models. Indonesia's ability to withstand the 2008 global financial and food crises is a case-in-point. In Indonesia, increased agricultural production, successful price stabilization at the consumer level, and trade controls thwarted the upward pressure on prices that might otherwise have been expected due to instability of rice supply in the world market. Despite prices doubling in international markets, domestic prices increased by only ten per cent. Indonesia achieved international recognition due to these relatively stable prices, with its national rice stocks remaining affordable. Certainly, Indonesia was able to avoid the food riots that occurred in other parts of Asia, Africa and Latin America. In this situation, it was possible for the government to maintain sufficient buffer stocks of rice due to good levels of domestic production, thereby preventing further speculation or hoarding.

To achieve this, the Indonesian government took a number of measures to limit the domestic impacts of the 2008 food crisis (taken from FAO, 2009). These included: removing import duties on wheat and soybean imports; providing soybean subsidies to producers of tofu and *tempe*; relaxing the VAT (value-added tax) for wheat flour and cooking oil; and increasing the fertilizer subsidy by 240 per cent. In terms of food accessibility, the government increased rice subsidies through RASKIN both in terms of programme coverage and the size of the subsidy, and introduced direct cash subsidies for the poor. This mix of policies and financial incentives provided household-level benefits, but overall Indonesia's relatively strong food security in 2008 was built on the level of price control made possible through solid domestic production.

The global food crisis of 2008 has been interpreted by many as an early warning sign of global economic conditions in the years ahead, as Indonesia and other countries are expected to grapple with the vagaries of climate change and its impacts on food production. In this context, having national control over the life source of the nation holds both emotive and practical appeal. The collective national memory of Indonesia's inability to buy food on the world market in the 1960s (and hence the perceived need to be self-sufficient) is juxtaposed over the experiences of localized food shortages and famine (such as those in Yahukimo) caused by entrapment in a subsistence economy and isolation from broader trade and social networks. These two typified experiences, and the complexities of addressing food insecurity as multi-scalar phenomena, are driving what may seem to be contradictory policies at national and regional levels. Singularly prescriptive policy approaches such as 'self-sufficiency at all costs' or 'full liberalization of agricultural markets' are unlikely to solve the food security conundrum in Indonesia. Food security is most likely to be achieved through broad-based rural development, where individual households are supported to improve their access to resources and factors of production through an entitlements approach.

## References

AHRC (2009) 'Indonesia: 113 villagers' hunger deaths caused by government neglect as well as harvest failure in Yahukimo, Papua', *Asian Human Rights Commission (AHRC) – Hunger Alert Programme*, 16 September 2009, www.ahrchk.net/ua/mainfile.php/2008/3260/, accessed December 17, 2010

Arifin, B. (2009) *The Role of the Private Sector in Indonesia's Rice Industry*, report submitted to the ASEAN Secretariat and the World Bank. October 31, 2009

Bappenas (2007) *Report on the Achievement of Millennium Development Goals Indonesia 2007*, National Development Planning Agency (Bappenas), Jakarta

BPS (2010) Service Centre, Badan Pusat Statistik (BPS, Central Statistics Agency), www.bps.co.id

Deptan (2010) 'Mentan Grandlaunching Food Estate Merauke', *Berita Pertanian Online*, Indonesian Department of Agriculture (Deptan), www.deptan.go.id/news

DKP (2010) *Food Security and Vulnerability Atlas* (FSVA) of Indonesia, jointly released by the World Food Program and the National Food Security Agency (DKP), Jakarta, http://foodsecurityatlas.org/idn/country/fsva-2009

Fane, G. and Warr, P. (2008) 'Agricultural Protection in Indonesia', *Bulletin of Indonesian Economic Studies*, vol 44, no 1, pp133–50

Fane, G. and Warr, P (2009) 'Indonesia', in L. Anderson and W. Martin (eds) *Distortions to Agricultural Incentives in Asia*, The World Bank, Washington, DC, pp165–196

FAO (2009) *Measures taken by governments to limit the impact of the Global Food Crisis*, FAO/GIEWS, FAO, Rome

FAO (2010) *The State of Food Insecurity in the World: Addressing Food Insecurity in Protracted Crises*, FAO, Rome.

Hajramurni, A. (2008) 'Another baby dies from malnutrition in Makassar', *The Jakarta Post*, 12 February, 2008

McCulloch, N. (2008) 'Rice prices and poverty in Indonesia', *Bulletin of Indonesian Economic Studies*, vol 44, pp45–64

McCulloch, N. and Timmer, P. (2008) 'Rice policy in Indonesia: a special issue', *Bulletin of Indonesian Economic Studies, vol* 44, pp34–44

Martianto, D. (2009) *Indonesia Tahan Pangan dan Gizi 2015* (Achieving Food and Nutrition Security of Indonesia in 2015), draft version, Agency for Food Security, Ministry of Agriculture, Jakarta

Neilson, J., Arifin, B. Fujita, Y., Hartari, D. F. S. (2011) 'Quality upgrading in speciality coffee chains and smaller livelihoods in Eastern Indonesia: opportunities and challenges', *Proceedings of the 23rd International Conference on Coffee Science*, Bali, October 3–8, pp454–462

Rakyat Merdeka (2010) 'Merauke Food and Energy Estate Mungkinkan Konflik Berkepanjangan di Papua', 6 September 2010, www.walhi.or.id/en/media-space/1824-merauke-food-and-energy-estate-mungkinkan-konflik-berkepanjangan-di-papua

Rigg, J. (2006) 'Land, Farming, Livelihoods, and Poverty: Rethinking the Links in the Rural South', *World Development*, vol 34, no 1, pp180–202

Ruf, F. and Yoddang (2001) 'Cocoa farmers from boom to bust', in F. Gerard and F. Ruf (eds) *Agriculture in Crisis: People, Commodities and Natural Resources in Indonesia, 1996–2000,* Curzon Press, Richmond, pp97–156

Sen, A. (1981) *Poverty and famines: an essay on entitlement and deprivation*, Clarendon Press, Oxford

Setneg (2010) 'Food Estate, Harapan Ketahanan Pangan Masa Depan', *Online News*, 25 February 2010, State Secretariat of the Republic of Indonesia (Setneg), www.setneg. go.id

Siregar, H. (2007) 'Agricultural Development in Indonesia: Current Problems, Issues, and Policies', paper presented at FAO-SEARCA Policy Workshop, *Asian Economic Renaissance: Challenges and Consequences on Agriculture, Food Security, and Poverty*, in Chiang Mai, Thailand, 19–20 March 2007

SMERU (2008) *The Effectiveness of the Raskin Program*, Research Report, The SMERU Research Institute, Jakarta, www.smeru.or.id/report/research/raskin2007/raskin2007_ eng.pdf

SPI (2008) *Pandangan dan Sikap SPI tentang kedaulatan pangan (SPI's policy stance and approach to food sovereignty)*, accepted in Malang, East Java on 28 February 2003, posted by Serikat Petani Indonesia (SPI, Indonesian Peasants Association) on 2 January 2008, www.spi.or.id, accessed 22 December, 2010

Thomas, M. and Orden, D. (2004) *Agricultural Policies in Indonesia: Producer Support Estimates 1985–2003*, MTI Discussion paper no 78, Markets, Trade, and Institutions Division of the International Food Policy Research Institute (IFPRI), Washington, DC

*Time Magazine* (1964) 'Indonesia: of rice and rats', *Time Magazine*, 28 February, 1964

Timmer, C. P. (1996) 'Does BULOG Stabilize Rice Prices in Indonesia? Should it Try?', *Bulletin of Indonesian Economic Studies*, vol 32, no 2, pp45–74

Timmer, C. P. (2004) 'The Road to Pro-poor growth: the Indonesian experience in regional perspective, Bulletin of Indonesian Economic Studies, vol 40, no 2, pp177–207

Warr, P. (2005) 'Food Policy and Poverty in Indonesia: A General equilibrium Analysis', *Australian Journal of Agricultural and Resource Economics*, vol 49, no 4, pp429–51

World Bank (2005) 'Feeding Indonesia', *Indonesia Policy Brief*, www.worldbank.or.id, accessed 20 December 2010

# 11

# 'SOYIZATION' AND FOOD SECURITY IN SOUTH AMERICA

## *Navé Wald, Christopher Rosin and Doug Hill*

In the Southern Cone of South America, the cultivation of soybeans embodies many of the issues related to existing conceptions of the global food system outlined in the first section of the book. Soybeans – as an integral component of meat production systems in Europe and China (see Campbell, this volume) as well as more recent contributions to the promotion of biofuel (see McMichael, this volume) – represent an important source of export income for the debt-ridden economies of Argentina, Brazil and Paraguay (and more recently Uruguay and Bolivia). In addition to meeting the financial needs of these countries, soybeans facilitate their participation in international trade and the neoliberal 'solution' to food security (see Pritchard, this volume, and compare to Neilson and Arifin, this volume). Because cultivation of soybeans largely involves 'modern' management practices including mechanization, advanced no-till cultivation and GMO (Genetically Modified Organism) seeds, it has been readily incorporated within the efforts of the countries to modernize and conform to the quantity idealism of the global food system (see Stock and Carolan, this volume).

As noted in the companion chapters in the second section of this book, integration to the global food system both has profound effects on and elicits creative response from local food producers. For those accustomed to an agricultural sector in which soybeans are part of a patchwork of industrialized commodity production alongside maize or wheat, the impact of the crop on landscapes and livelihoods (referred to as *soyization*) in the Southern Cone of South America is starkly evident. In Argentina, Brazil and Paraguay, rather than involving the introduction of an alternative cash crop in highly mechanized farming systems, soybean cultivation imposes significant land use and socio-economic changes on either forested lands or small-farming (*campesino*) landscapes.

These changes have not occurred without the resistance and mobilization of rural social actors. Livelihood challenges associated with neoliberal commodity production have enhanced the consolidation of social movements across these regions, which continuously contest this agro-export model. Such notable movements are the Brazilian Landless Rural Workers' Movement (*Movimento sem*

*Terra* – MST) and the Paraguayan National Coordinating Board of Peasant Organizations (MCNOC). In this chapter we highlight some of the main socio-economic effects of the soyization process in Argentina, Paraguay and Brazil, and use an Argentinean case study of MOCASE-VC (*Movimiento Campesino de Santiago del Estero – Vía Campesina,*[1] or Peasant Movement of Santiago del Estero – *Vía Campesina*) as an exemplary anti-hegemonic peasant organization to show the types of local initiatives that have emerged in response to expanding soybean cultivation.

## The advancing soybean frontier and its impacts

For the authors, our engagements with the people occupying the landscapes affected by soy production also evoke insight to the relationship between commodity soybean production and global food security. One of us (Chris) has experienced the impact of soybean expansion as a temporal development. On an initial trip to the Eastern Paraguayan province of Alto Parana in 1990, he recalls driving a recently sealed highway through expanses of semi-tropical forest. The wall of trees bordering the road was broken occasionally by encampments of land-less peasants that encroached no more than 50 meters into the dense tree cover. More developed settlements of immigrants (from Brazil, Europe, North America or Japan) were only evident in road signs pointing to a *Colonia* located to the east, nearer to the Parana River. Subsequent trips along the same road in 1996 and 1999 revealed a landscape markedly devoid of trees and with cultivated land stretching to the horizon. The landless encampments had been replaced by isolated clusters of smallholders' homes, the occasional bustling town and several immigrant settle-ments distinguished by larger houses and abundant agricultural machinery. The closed, other-worldly confines of the forest had been completely replaced by vast, sparsely-populated swaths of industrialized agriculture. In this landscape, even the smallholders often turned to soybeans – grown with machinery contracted through larger landowners – as the only economically viable agricultural activity. Due to the scale of production imposed by the machinery employed, the soybeans have replaced much of the mixed subsistence and cash (formerly cotton) cropping of traditional Paraguayan smallholders.

For another of the authors (Navé), the impact of soybeans was evident in the spatial contrast between native forest and cultivated fields, where the 'agriculture frontier' is stretching further north from the fertile plains of the Pampa into the semi-arid Gran Chaco region. This process is highly evident in the Argentinean province of Santiago del Estero, where native forest is being gradually cleared for cultivation. In some parts of the province, and particularly in the east, formerly dense forest is now depleted of its valuable timber and often reduced to mere corridors or islands of bush around and within large cultivated fields. This process of deforestation is forcing peasant families who depend on the forest for their subsistence to relocate, resulting in the expansion of small towns and communi-ties along the main roads. Also increasingly evident, and in striking difference to the peasants' adobe houses along the main roads of the province, are *estancias*, which often include a modern brick house and sheds for heavy agricultural machinery. In Argentina, this type of farmhouse has been traditionally associated with the farming model of the pampa. The role of soybeans in this expansion is reflected in the numerous granaries, many owned by Cargill, that can be found along the main roads and in rural towns.

Beyond the visual impacts of soybeans on the landscape, it is easy to find statistical evidence of the emerging dominance of soybeans in the agricultural sectors of Argentina, Brazil and Paraguay. The countries currently rank as three of the four largest soybean exporters globally, demonstrating the value of the crop for economies dependent on export earnings. Each has also seen increasing areas under soybean cultivation during the last two decades: recent figures have Brazil reporting 22 million hectares, Argentina 16.6 million (increasing from 5.1 million in 1989–90) and Paraguay 2.6 million (increasing from 800,000 in 1994–95) (cited in Altieri and Pengue, 2006). It is perhaps yet more telling that the figures for the latter two countries respectively represent 50 and 80 per cent of their total cultivated land. Soybeans place ever-increasing pressures on already disparate land distributions and relatively insecure land rights regimes. In Paraguay, for example, less than 2.5 per cent of landowners own over 85 per cent of the land, while 41 per cent claim areas of five hectares or less (Barreto Monzón, 2009). In Argentina, the expansion of soybean cultivation is associated with a 21 per cent reduction in the total number of farms between 1988 and 2002 (Aizen *et al.*, 2009). In Brazil, similar tendencies have been documented for the exclusion of smallholder agriculture where soybean cultivation is experiencing rapid expansion in the Cerrado and Amazon regions (for example, Altieri and Pengue, 2006). These conditions reflect the local impact of the global promotion of biofuels and trade policies that promote commodity production.

Despite its contribution to the global food system and the extent of landscape change for which it is responsible, awareness of the impacts of soybean cultivation in South America is not widespread. Environmental NGOs (non-governmental organizations) such as Greenpeace and the World Wildlife Fund, however, have used aerial photographs of forest clearance in the Amazon and the border areas of Argentina, Brazil and Paraguay surrounding the Iguazu Falls as graphic depictions to draw international attention to its environmental impacts. While these campaigns have elicited efforts by food corporations (led by McDonald's in Europe) to eliminate soybeans produced at the expense of tropical forest clearance from their supply chains, they have also been dampened by a focus on promoting 'sustainable soy' by these very same organizations (Steward, 2007). The latter actions imply that the negative (environmental) impacts of soybean cultivation can be contained through compliance with best practice standards. The social implications of expanding soybean cultivation receive even less media attention outside the region, with only occasional stories of glyphosate poisoning (associated with the use of Roundup Ready (RR) soybeans) and forced land appropriations appearing in national and regional newspapers. Even less well publicised are the impacts of the soybean frontier on global food security and the Food

Sovereignty[2] of the populations in the affected regions. These latter impacts are twofold, involving: 1) the dislocation of smallholders unable to 'compete' with commodity soybean production for access to land, and 2) the substitution of traditional food and staple crops by a product largely intended for animal feed or biofuel production outside the region. Together, these factors threaten *campesino* livelihoods to the extent that collective action has become a viable response – as is illustrated in the case study in the second part of this chapter.

The emergence of soybeans as a dominant and highly-valued crop in the Southern Cone is tellingly emblematic of the economic development of the region. Prior to the 1980s, the focus on import substitution policies and the desire to modernize both societies and economies led to large investments in industrial infrastructure, especially in Brazil and Argentina. Such investments relied heavily on the availability of international finance, fed, in particular, by the flush of capital accruing in petroleum producing countries (Kay, 2004). This strategy realized mixed development results, and eventually led to severe indebtedness subject to sharp interest rate increases during the 1980s. In the aftermath, the International Monetary Fund and the World Bank were able to leverage neoliberal reforms (coined 'structural adjustments') across the region, a process which included privatization of state enterprises and a sudden decline in government social expenditure. This restructuring diverted the economic development focus from industry to agriculture, in which the region was deemed to have a comparative advantage, and forced small-scale farmers and peasants to compete with better resourced commercial, and often multinational, agribusiness (Kay, 2004; Porto, 2007). The associated imperative to reduce national debt levels also meant that the focus of agricultural production shifted from meeting domestic market demands to promoting crops with the most lucrative export potential – a role which soybeans would eventually fill.

In addition to fitting well within the emerging political economic arena of late 20th century South America, soybeans embodied the means to achieve a long-held desire to modernize agricultural production. The marginal value of soybean production limits its viability as a cash crop, except where large areas can be managed by a relatively small amount of labour. In other words, soybeans only prove profitable as a commodity, under mechanized production. From its introduction to southern Brazil, the crop has involved the replacement of manual with mechanized labour in agricultural production (Banck and den Boer, 1991). With increasing value placed on larger areas of cultivation – considered more efficient for mechanized use – rural populations were displaced as share-cropping and similar land leasing arrangements become less tenable and small landholders were pressured to sell or abandon their farms. The emphasis on modernization has also enabled the introduction of GMO seeds in South America, led by RR soybeans, which facilitate no-till management and further reduce running costs of farm machinery while simultaneously increasing reliance on herbicides.

The process through which soyization has impinged on the land access rights of local *campesinos* while privileging the production of an industrialized commodity offers a dramatic example of challenges that a global food system based on the neoliberal model poses to Food Sovereignty. Philip McMichael (1997) attributed the problem of food dependency in the developing world to the green revolution of the 1970s, which incorporated the agro-industrial model into food production in the South. Modernization of agriculture created dependency on imported inputs and enhanced the integration of the food market. Since the 1980s, it was increasingly notable that this North-South integration had initiated substitution of traditional basic food crops with commercial non-traditional feed crops and other high-value commodities, such as fresh fruit and vegetables, for export from developing countries. This shift was a response to an increasing demand for animal proteins and other high-value foods, driven by the proliferation of affluent diets primarily, but not exclusively, in the developed world (McMichael, 1997; McGlade, 1997).

According to Walter Pengue (2005, p319) 'we are facing a battle for high quality protein between developed and developing countries'. In the Southern Cone this shift has been experienced when a more diverse food production model is altered and replaced by the extensive cultivation of feed crops for animals, largely destined for Europe and China. As a result, poor people can no longer produce or afford the diverse diets they once enjoyed: traditional high-value meat protein grown on less intensive pasture has been displaced by vegetable protein, such as soybeans. This nutritional change entails a cultural shift as well as adverse effects on health (Pengue, 2005). Farshad Araghi (2000, p155) refers to this phenomenon as the 'hunger amidst abundance' that materializes when agriculture is directed towards affluent diets and a process of depeasantization occurs in which the rural poor lose the means to subsist.

At the broader scale, the soyization of Southern Cone agriculture appears as a relentless tide through which the peasantry is being disadvantaged in a global food market largely controlled by powerful agro-multinational corporations (McMichael, 2000). *Campesinos*, on one hand, cannot afford to modernize in order to compete in the global market; while on the other hand, their production of basic foodstuffs is challenged by loss of land to large farmers and by imports of artificially cheap foodstuffs produced by subsidized farmers in the North (McMichael, 1997). That, in turn, has led in some cases to food deficiency and dependency, as well as malnutrition (Pengue, 2005). It is, however, possible to identify actions of *campesino* organizations that seek to contest the incursion of soybean cultivation. While it is not possible at this point to argue that such activities are successful (the area of soybean cultivation continues to grow), we believe the existence (and persistence) of these organizations offers an indication of the potential to promote and improve the type of small-scale practices and technologies that are posed as more appropriate solutions to global food insecurity (see Godfray *et al.*, 2010). The MOCASE-VC organization offers insight to both the potential of, and challenges to, locally initiated attempts to contest the expansion of commodity agriculture.

# Local response to soyization in Santiago del Estero, Argentina

In the Southern Cone, soybean cultivation has left an indelible mark on the land-scape and on the social conditions of its inhabitants. The province of Santiago del Estero in northern Argentina (see Figure 11.1) is among the areas most recently impacted by soyization and, given Navé's current research in the region, we use it as a case study to highlight some of the real-life implications of the process. While the particular social, political and environmental conditions are unique to the region, the impacts on land tenure, human and animal health, and productive capacity are common to areas of soybean cultivation throughout eastern Paraguay and in southern and central Brazil. The response of the *campesino* groups in Santiago del Estero is also somewhat distinctive in the scale and coordination of the membership. Their aspirations for appropriate technology, community-led development and active contribution to domestic food security resonate with those of small-scale producers encountered elsewhere in the Southern Cone. Thus, the case study provides insight to both the challenges posed by soybean cultivation as well as the potential, locally-initiated solutions that correspond to greater nutritional security and environmental sustainability.

## *The expanding agriculture frontier and the campesinos in Santiago del Estero*

Santiago del Estero is Argentina's least urbanized province and one of its poorest. Compared with other parts of the country, it is marked by a somewhat different process of colonization in which its rural population stayed *criolla* – of indigenous and European lineage – comprising, to a large extent, of *campesinos*, or peas-ants (Durand, 2009). While historically highly marginalized in political, social and economic terms, the peasant sector is relatively important to the province's economic capacity (de Dios, 2006). Although official statistics do not provide explicit information about this sector, the subcategories of the census category of 'small producers' can be used to separate peasant from more capitalized small producers, and thus serve as a proxy for the extent of the *campesino* sector. This analysis suggests that *campesinos* account for approximately 83 per cent of total farms, although a mere 16 per cent of the total agricultural land in the province (Obschatko *et al.*, 2007). Despite their representation in the agricultural census and historical claims to specific lands (at times for generations), they largely occupy land without having legal recognition as proprietors (de Dios, 2009). Insecure land tenure remains one of the main problems confronting the *campesinos* in the province, especially as it is considered a likely region for soybean cultivation.

The precarious land tenure rights have made *campesino* farms a soft and frequent target of resourceful agribusinesses wanting to expand their operations, even prior to the emergence of soybeans, when the agribusiness was primarily focused on cattle, cotton and maize. The first land dispute resulting in the eviction

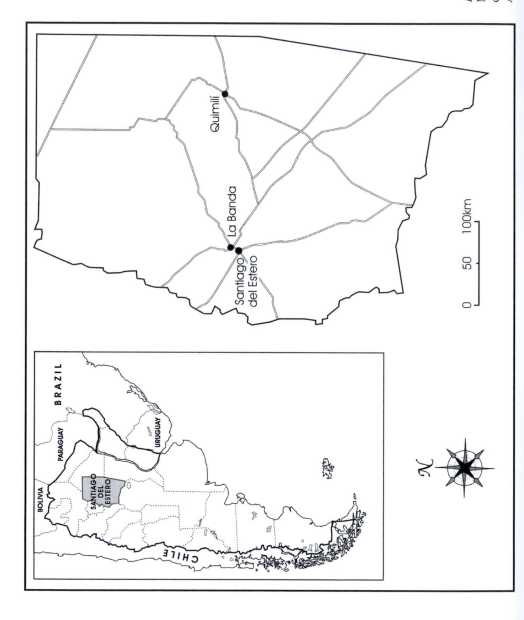

*Figure 11.1*
Map of Santiago del Estero Province, Argentina

of *campesinos* occurred in 1963, in the southeast of the province. The 1970s saw an increase in the numbers of judicial evictions and displacements of *campesinos* when a number of commercial agriculture companies claimed ownership over parcels of land already occupied by *campesinos* (Barbetta, 2009).

In 1986 another dispute over land erupted when a number of companies intended to evict some 400 families of *campesinos* occupying about 120,000 hectares in the eastern part of the province. With the encouragement and support of a local Catholic clergyman and an NGO, the *campesinos* organized to successfully defend their land (Durand, 2009). Following this event, further mobilization of the peasantry led to the establishment of a number of organizations across the province. In 1990 these organizations joined forces to create the *Movimiento Campesino de Santiago del Estero* (Peasant Movement of Santiago del Estero – MOCASE). In 2001 an internal dispute led to a separation within the movement that resulted in the formation of two organizations – MOCASE and MOCASE-Vía Campesina (VC). The main difference between the two organizations involves their internal and external politics: their principal objectives remain virtually the same – that is, securing land tenure and improving standards of living of their communities. As in other parts of Latin America, insecure land tenure is at the centre of their mobilization, although the struggle also reaches much further to address social and economic objectives. The MOCASE-VC struggle for secure tenure involves more than the land's economic potential commonly associated with property rights. For *campesino* communities, some of which are of indigenous identity, land is not an economic asset, it is life: 'The most serious challenge that we have is to defend the land. And defending the land means defending life, the life of our *Pachamama*, because the land is our mother', (Informant 1, MOCASE-VC, interviewed 22.2.2010).

The struggle for land is also about the historical marginalization of the *campesino* sector both in the province and nationally. It is about political exclusion, atrocities, a predatory economic environment, lack of services and cultural intolerance by the hegemonic economic and political elite. Equally, it is about human and citizen rights, food sovereignty and autonomy.

Since the late 1980s, the mobilization and organization of the *campesinos* in Santiago del Estero has aimed to contest the political culture and economic models that disadvantage them. Because *campesinos*' livelihoods are rooted in their land both economically and socially, the issue of land tenure is paramount for them. As one member of the MOCASE-VC asserted:

> The most severe challenge is that if we want to be free we must begin to produce food, produce our own food … And one of the most serious challenges for this is that the spaces and the land [that *campesinos* have] are very small spaces that do not allow the development of the family, as a natural development.
>
> Focus Group 2, MOCASE-VC, interviewed 20.3.2010

An important notion here is the moral claim to the freely chosen ability to produce one's own food, to avoid dependence on others. Perhaps of greater importance, though, is the multifaceted significance of land. Loss of land entails not only production-related ramifications but also serves as an impediment to the social reproduction of the *campesino* family unit.

Traditionally, *campesino* families in Santiago del Estero cultivate small and private parcels of land while their animals graze freely in the forest, which is treated as a common, alongside the herds of neighbouring families. The advancement of the agriculture frontier has initiated a process of enclosure. As a result, these commons have diminished in area (in some parts of the province more than others) to a point where they cannot sustain the *campesinos'* herds, let alone the natural reproduction of their family. If space constraints do not allow for social reproduction of the peasantry, then a process of depeasantization, where peasants become proletariats – as theorized in the Disappearance Thesis (Araghi, 1995) – is inescapable. In the process, the benefits of diversified production to the diets of the *campesinos,* as well as the knowledge that enables their successful exploitation of seasonally dry landscape, will be lost.

Therefore, in order to allow for their livelihoods to prosper, it is crucial for the *campesinos* to have access to sufficiently large parcels of land. Beyond the importance of land tenure, however, lie challenges to the economic viability of peasant agriculture production. Here, again, modern intensive agriculture practices are adversely affecting the *campesinos'* livelihoods, in terms of both subsistence farming and petty trade of surplus production. A MOCASE-VC member explains:

> A greater challenge that we have is to produce, and of great concern to us is the issue of transgenic crops that make you fail year after year, and one can denounce [it to the authorities but] the judiciary does nothing, only reacts for the system, for the agro-exporting model, which is installed for about 20 years here in the area where we live, which is an area that for many years was not being taken into account because [the soils] were not fertile. But since these new genetically modified seeds [arrived] soils are fertile, and today they come for territory, for the land, and they don't care about the disasters they cause, like knocking down millions of hectares of woods, forest, native forest.
>
> Focus Group 2, MOCASE-VC, interviewed 20.3.2010

In only a few sentences, this *campesino*-turned-activist captures much of the essence of the *campesinos'* struggle in Santiago del Estero above and beyond the issue of land tenure. That is, he emphasizes the advancement of the agriculture frontier into previously marginal land, the severe consequences of this process for local people and the protection the agribusiness sector has from the authorities. He then added that:

What we have to do is to expel all that soybeans system, this agribusiness, and start producing for the market; producing healthy foodstuffs. We are capable of producing food, but what happens is that the system limits you. First is the lack of land, the territory that they take away from you. Second, the food market which is controlled by big multinationals that prevent you from putting a product on the market.

Focus Group 2, MOCASE-VC, interviewed 20.3.2010

That is, the path toward sustainable development for the *campesinos* in Santiago del Estero is simple: producing organic foodstuffs for the market. However, while the issue of land tenure and its importance for peasant livelihoods is clear, the second point here is more often overlooked. Provision of products for the formal market is being increasingly restricted and regulated. While the public health reasoning for such restrictions and requirements for certification of foodstuffs is well understood, many *campesinos* feel that this system disadvantages them, restricting them from selling their products on the market. For example, some communities of the MOCASE-VC produce pickled goat meat for sale in local and regional markets. This kind of processed food requires certification in order to be commercialized. However, there is only one slaughterhouse that is permitted to kill small farm animals (goats, sheep, etc.), and it is located far to the south of the province – which means transportation costs are prohibitive for *campesinos* located elsewhere in the province. Therefore, if *campesinos* want to process the meat of these animals, as they do, they cannot possibly obtain the required certifications.

## The health implications of soybean cultivation

Another important aspect of the soyization process is the intensive application of herbicides from airplanes onto large fields of herbicide-resistant RR soybeans. In some parts of the province, *campesinos*, as well as small-town residents, live in close proximity to such fields and are thus being affected from residues of agro-chemicals being sprayed from the air. A MOCASE-VC activist asserted that:

The expansion of the agribusiness was not only evicting people in the sense of the violence of expulsion off the land, but daily contaminating their crops, contaminating their water sources, [even] poisoning the people themselves when the airplanes are spraying.

Informant 2, MOCASE-VC, interviewed 25.2.2010

The use of agrochemicals, which has intensified, particularly outside the Pampa region, with the expansion of the soybeans model, constitutes a serious threat to *campesino* food production. According to another activist of the organization:

There are families who have nothing left to produce with because they have been either sprayed by glyphosate [herbicide], or this poison was thrown into the water sources and all the animals have died, [and then] they have to start over a life that they have already had before, and those things have caused a setback in the production.

Informant 3, MOCASE-VC, interview 18.3.2010

This is not only an issue when the wind often carries the agrochemicals onto neighbouring *campesinos'* fields and destroys crops or poisons animals: there are also testimonies of airplanes spraying houses directly. The contamination of water sources, as mentioned above, is having a far-reaching effect on both livelihoods and health in the countryside.

Since many countryside dwellers collect rain water from their rooftops, agrochemicals make their way into drinking water supplies, contaminating them and affecting people's health. Rural schools are no exception. In some parts of the province, where soybeans cultivation is prevalent, there are schools completely surrounded by fields of soybeans and the children who attend them drink storm water collected from the roofs of their classrooms. It is a somewhat dramatic depiction of the current proliferation of modern agriculture in northern Argentina, but that is reality for many rural dwellers.

There are regulations in place that prohibit spraying from the air in proximity to people's houses but, as hinted above, the authorities do not necessarily act to enforce them. This is a key limiting factor on efforts to limit the proliferation of the agriculture frontier and its associated costs. Namely, laws that are designed to protect the people and the environment exist but enforcement is nominal. Without the work of diverse peasant organizations and civil groups, this situation would be even worse. The inevitable ramification of this reality is that owners of large farms can expect few, or no, consequences for breaching laws and regulations, especially in remote areas.

### Campesino definitions of poverty

A noteworthy aspect of the contesting of soyization in Santiago del Estero is the lack of direct association with food security among the *campesinos*. Notwithstanding the challenges of production mentioned by members of the MOCASE-VC as having a significant effect on their livelihoods – including the harsh climatic conditions and recent droughts as well as the soybeans and agrochemicals – virtually no one referred to either food insecurity or shortage. While conversing one night with a member of MOCASE-VC, he asked if there are poor people in Navé's country. This question led to a discussion on what defines poverty and who is poor. Navé then asked if he was poor and he responded 'well, no'. He then explained that he has a roof over his head and is not worried about the next meal; his animals and produce provide him with most of the food he consumes. This is not an unusual response regarding notions of who is poor and what constitutes poverty. Food insecurity is definitely associated with poverty and some argued that one would find it in urban slums but not in the Santiagueño countryside. Be that as it may, *campesinos* that were impacted more severely by a variety of challenges, including those discussed above, had migrated into urban settlements where food insecurity may be greater.

One reason urban dwellers perceive the *campesinos* to be poor, however, involves their basic living conditions. Although this MOCASE-VC member is content with the material comfort of his home, which has no running water or sewage and only a 12-volt solar panel for electricity, he admitted that if someone would offer to pay for the installation of a modern bath and toilet he would not resist having it. Like other *campesinos* with whom we have spoken in Argentina, Brazil and Paraguay, he does not resist modernity. He readily acknowledges, however, that goods and services such as satellite television or internet mean more and more bills, just like urban dwellers have, and *campesinos* more often than not lack disposable money for such amenities.

It is widely accepted among MOCASE-VC members that modernizing some production methods, such as having a tractor or a chainsaw, will be beneficial. Thus, although they contest soyization as a modernization process, the *campesinos* do not reject modernity entirely. Rather, many of them want technology that is appropriate for their way of life. For instance, they reject technology that reduces the need for labour and eliminates the means for people to make a living. Modernization and technology are not necessarily perceived as bad; but they are seen as potentially changing the aspirations of the younger generation in ways that could challenge existing *campesino* livelihoods and culture. Therefore, technology is welcomed as long as it is useful and practical to the *campesinos*. What exactly is useful and practical, however, may be debateable.

### *Appropriate technology and development*

The potential for a small tractor and associated farming implements to reduce the difficulties of cultivation as well as improve the capacity for timely management interventions is an example of a desired technology. The latter capacity is of particular importance in a challenging environment where crop performance is often limited by available soil moisture. The introduction of such technology is further conceived as a common good, with shared access across the community. A member from the south of the province, who expressed reservations regarding having electricity, had a different perspective in regard to having a tractor:

> Here what the people are always talking and saying is that we want to have our own tractor in the community with all the equipment to allow us to plough, to allow our mates that are committed to plough for all the comrades, because a lot of people would like to sow, and, well, what they sow is with a plough, with a handlebar, harnessed to a horse. But they sow a little bit, more for [self]-consumption.
>
> Informant 4, MOCASE-VC, interviewed 27.3.2010

Some communities of the MOCASE-VC do own a tractor, but not all of them are able to purchase such equipment. Lack of appropriate credit lines for purchasing and maintaining even small-scale machinery compels many communities to rely on traditional cultivation methods and this, in turn, constrains the scale of their production.

The aspirations of the *campesinos* for land, affordable credit, appropriate machinery and the infrastructure and services that would assist them to improve production and to provision both the domestic market and themselves with healthy foodstuffs, are all articulated within their demand for an integral agrarian reform. In their own words:

> It is called an integral agrarian reform; not only to say 'come along, there you have land and go for it'. [It's also about] infrastructure, training. If I give you 50 hectares of land and I tell you, 'you live there!' how would you live? You will die of hunger because you have to have animals, knowledge, you have to have a cistern of water, you have to have a ranch [house], you have to have chickens. Then, you have to know, to learn, how to make an organic vegetable garden, how to raise animals. And that is an agrarian reform ... and not that it would be an agrarian reform imposed by the government: "this is the agrarian reform!" They should come and sit with us, those who really live [off the land and they should ask us] what do we want? How do we want to manage it? How do we want to implement it? That would be an agrarian reform; helping you with production, with commercialization.
>
> Informant 5, MOCASE-VC, interviewed 1.3.2010

Notwithstanding the local livelihood challenges and the aspiration for an integral agrarian reform, the counter-hegemonic discourse of the MOCASE-VC reaches far beyond the local or the national. It is well recognized that the soyization process is part of a global economic system that is not only unsustainable but also disastrous. Commenting on the global food system and the role peasants *should* have in it, a member of the MOCASE-VC asserted that:

> In my opinion, the one who would be producing foodstuffs in large quantity would be the *campesino*. Healthy foodstuffs ... a sustainable production. A time will come when the small *campesino* will be the one who provides food to the world. Here they come and sow soybeans. In two or three years of sowing soybeans, they won't have fertile land for further cultivation; they will stop sowing soybeans. And that [land] is going to be left behind and they won't produce anything, because they are not going to be able to produce. In contrast, the economy of the *campesino* is smaller but more reliable.
>
> Informant 5, MOCASE-VC, interviewed 1.3.2010

The vision here is for a system of food production driven by small producers, who will produce primarily for human consumption and with greater environmental consideration and responsibility. Currently, small producers, such as the *campesinos*, are facing great challenges not only to produce foodstuffs for the market but also for sustaining their own families.

The climatic conditions in Santiago del Estero are not favourable for many crops but this is primarily due to scarcity of water. Investments in water infrastructure followed by adequate management would allow many *campesinos* to increase both the area of cultivation and their productivity. Equally, availability of small-scale machinery would further contribute to productivity. However, investments in rural infrastructure often raise land values. For this reason many communities struggle for communal land titles that would make it extremely difficult for resourceful buyers to purchase land once transferable titles are issued. Under such conditions *campesinos* could prosper and contribute even more to food security.

## Conclusion

As the case study of Santiago del Estero demonstrates, soyization is only the latest chapter in an ongoing struggle among small-scale land managers for security of land tenure and the capacity to produce diverse foods as an element of food sovereignty in the Southern Cone. The historical marginalization of *campesinos* has established the conditions under which not only families but entire communities are forced off their land and the surrounding native forest is cleared. Due to the domestic focus on the achievements of modernization and the focus of international concern on, specifically, environmental impacts, the fate of *campesinos* has largely remained a 'concealed' aspect of the advancing agriculture frontier; an aspect that, prior to social mobilization, went publicly unreported and thus unnoticed. It becomes increasingly evident that the expansion of the agriculture frontier, with its fences, monoculture of soybeans, intensive use of agrochemicals and deforestation are all vivid threats to the livelihoods of *campesino* families, to the environment and to global food security.

On the other hand, soyization also provides a focal point for *campesino* resistance that is fertile ground for the evolution of local 'solutions' to the social and environmental dangers of commodity soybean production. Assisting *campesinos* – and other small-scale farmers – not merely to survive but to prosper, could have far-reaching implications for global food production and security. Small- and medium-size producers, it is argued, could feed the growing global population in a more sustainable manner if given the right conditions (Godfray *et al.*, 2010). For creating these conditions, however, governments must take an active role, even if it is merely enforcing the law. In Argentina, for example, it needs to halt the advancement of the agriculture frontier and to enforce existing regulations regarding deforestation and use of agrochemicals. Further progress would be achieved through a *genuine* process of consultation and collaboration, with the objective of creating an economic environment that accommodates the needs of small-scale producers. The implications of meaningful change extend beyond narrowly defined agrarian reform to include training, concessional credit, infrastructure (transportation, water and more) and other services. The result would challenge the existing reliance on commodity soybeans for animal feed in China and Europe, while providing the means for more diverse diets and food sovereignty in the Southern Cone.

## Notes

1   La Vía Campesina (The Peasant Way) is a transnational movement of peasants, small- to medium-size agriculture producers, landless agriculture workers, rural youth and women and indigenous people. Among its main objectives are social justice, fair economic distribution of wealth, agrarian reform and food sovereignty (defined below). The MOCASE-VC is a member of this international movement.
2   The term 'food sovereignty' was coined in the 1996 World Food Summit by the transnational peasant organization La Vía Campesina, of which the MOCASE-VC is a member. La Vía Campesina (2007, p1, emphasis in origin) defines food sovereignty as: the RIGHT of peoples, countries, and state unions to define their agricultural and food policy without the "dumping" of agricultural commodities into foreign countries. Food Sovereignty organizes food production and consumption according to the needs of local communities, giving priority to production for local consumption. Food Sovereignty includes the right to protect and regulate the national agricultural and livestock production and to shield the domestic market from the dumping of agricultural surpluses and low-price imports from other countries. Landless people, peasants, and small farmers must get access to land, water, and seed as well as productive resources and adequate public services. food sovereignty and sustainability are a higher priority than trade policies.

## References

Aizen, M. A., Garibaldi, L. A. and Dondo, M. (2009) 'Expansión de la soja y diversidad de la agricultura argentina', *Ecología Austral,* vol 19, no 1, pp45–54

Altieri, M. and Pengue, W. A. (2006) 'GM soybean: Latin America's new colonizer', Grain, Seedling, www.grain.org/seedling_files/seed-06-01-3.pdf, accessed 25 October 2010

Araghi, F. A. (1995) 'Global depeasantization, 1945–1990', *The Sociological Quarterly,* vol 36, no 2, pp337–368

Araghi, F. A. (2000) 'The great global enclosure of our times: peasants and the agrarian question at the end of the twentieth century', in F. Magdoff, J. B. Foster and F. H. Buttel (eds) *Hungry for Profit: The Agribusiness Threat to Farmers, Food, and the Environment*, Monthly Review Press, New York, pp.145–160.

Banck, G. A., and den Boer, K. (1991) *Sowing the Whirlwind: Soya Expansion and Social Change in Southern Brazil*, Centrum voor Studie en Documentatie van Latijns Amerika, Amsterdam

Barbetta, P. (2009) 'En los Bordes de los Jurídico. Conflictos por la Tenencia Legal de la Tierra en Santiago del Estero', unpublished Ph.D dissertation, Faculty of Philosophy, University of Buenos Aires

Barreto Monzón, M. (2009) 'Case: Impacts of soybean monoculture in Paraguay: the case of the Lote 8 community', in M. S. Emanuelli, J. Jonsén and S. Monsalve Suárez (eds) *Red Sugar, Green Deserts*, FIAN International, Halmstad Sweden, pp. 278–285

de Dios, R. (2006) 'Expansión agrícola y desarrollo local en Santiago del Estero', paper presented at the *7th Latin American Conference of Rural Sociology*, Quito, November

20–24, 2006, www.alasru.org/cdalasru2006/03%20GT%20Rub%C3%A9n%20de%20 Dios.pdf, accessed 4 April 2009

de Dios, R. (2009) 'Los campesinos santiagueños y su lucha por una sociedad diferente', *Paper Presented at the 1st National Congress on Social Protest, Collective Action and Social Movements*, March 30–31, 2009, Buenos Aires

Durand, P. B. (2009) *Desarrollo Rural y Organización Campesina en la Argentina: La Experiencia del Movimiento Campesino de Santiago del Estero*, LibrosEnRed, Buenos Aires

Godfray, H. C. J., Beddington, J. R., Crute, I. R., Haddad, L., Lawrence, D., Muir, J. F., Pretty, J., Robinson, S., Thomas, S. M., and Toulmin, C. (2010) 'Food security: the challenge of feeding 9 billion people', *Science,* vol 327, pp812–818

Kay, C. (2004) 'Rural livelihoods and peasant futures', in R. N. Gwynne and C. Kay (eds) *Latin America Transformed: Globalization and Modernity* (2nd ed.), Edward Arnold, London, pp232–250

La Vía Campesina (2007) 'The international peasant's voice', La Vía Campesina online, http://viacampesina.org/en/index.php?view=article&catid=27%3Awhat-is-la-via-campesina&id=332%3Athe-international-peasants-voice&format=pdf&option=com_content&Itemid=44, accessed 18 July 2010

McGlade, M. S. (1997) 'The potential of irrigated lands to reduce food dependency in Mexico', *Society & Natural Resources,* vol 10, no 3, pp329–339

McMichael, P. (1997) 'Rethinking globalization: the agrarian question revisited', *Review of International Political Economy,* vol 4, no 4, pp630–662

McMichael, P. (2000) 'Global food politics', in F. Magdoff, J. B. Foster and F. H. Buttel (eds) *Hungry for Profit: The Agribusiness Threat to Farmers, Food, and the Environment*, Monthly Review Press, New York, pp125–143

Obschatko, E. S., Foti, M. P. and Román, M. E. (2007) 'Los Pequeños Productores en la República Argentina: Importancia en la Producción Agropecuaria y en el Empleo en Base al Censo Nacional Agropecuario 2002', Ministry of Economy and Production, Secretary of Agriculture – PROINDER, www.proinder.gov.ar/Productos/Biblioteca/contenidos/ESTINV.10.Los%20peque%C3%B1os%20productores%20en%20la%20Rep%C3%BAblica%20Argentina.pdf, accessed 8 May 2010

Pengue, W. A. (2005) 'Transgenic crops in Argentina: the ecological and social debt', *Bulletin of Science Technology & Society,* vol 25, no 4, pp314–322

Porto, G. G. (2007) 'Globalization and poverty in Latin America: some channels and some evidence', *The World Economy,* vol 30, no 9, pp1430–1456

Steward, C. (2007) 'From colonization to "environmental soy": a case study of environmental and socio-economic valuation in the Amazon soy frontier', *Agriculture and Human Values,* vol 24, no 1, pp107–122

# 12

# NEGOTIATING ORGANIC, FAIR AND ETHICAL TRADE: LESSONS FROM SMALLHOLDERS IN UGANDA AND KENYA

*Kiah Smith and Kristen Lyons*

## Introduction

Global food networks are increasingly governed by transnational standards, accreditation and certification systems regulating social and environmental conditions at distant production sites. These 'ethical' accreditation systems have emerged out of concerns about inequalities within the global food system, the environmental effects of agriculture and the social effects of the globalization of food production and, more recently, the food, fuel and financial crises. Responses include mainstream supermarket/industry global sourcing codes and standards (such as GLOBALG.A.P – see p184), grass-roots networks and labelling schemes such as Fair Trade, 'Ethical Trade' based on labour codes of practice, and organic certification systems. In each of these regulated alternative food networks, smallholder farmers[1] in the global South are positioned as beneficiaries (Blowfield and Dolan, 2008).

By re-embedding social and ecological relations into agriculture and food production, organics, fair and ethical trade all potentially provide opportunities for disadvantaged farmers in the South to access Northern markets with horticultural (and other) products that reflect emerging definitions of quality, ethics and sustainability. Organic farming,[2] for example, is framed in terms of its contribution to food security, as well as environmental and social justice, which is reflected in environmental standards related to biodiversity, soil fertility and water conservation, as well as social justice criteria around labour relations and gender equity (Mutersbaugh, 2004). Meanwhile, fair and ethical trade focus on 'how historically exploitative producer-consumer chains can be refashioned around ideas of fairness and equality' (Raynolds, 2002, p405). Fair trade schemes seek to establish economic justice and transform South-North trade through producer empowerment and poverty alleviation, while ethical trade normally refers to minimum labour standards (and occasionally environmental standards) within an existing trade model (Barrientos and Dolan, 2006).

While undoubtedly offering some benefits – mainly in terms of creating legitimacy for environmentally and socially sensitive production of niche market food exports from the South – these alternative food networks have also been widely criticized for reproducing the values and priorities of Northern consumers, policymakers, supermarkets, governments and others, thus undermining the quality constructions of diverse producers and products (see Blowfield and Dolan, 2008; Dolan, 2010). Much of this literature posits producers in the South as less powerful than other actors embedded in global food supply chains. These inequitable relations are constructed and maintained through certification systems.

This chapter presents a more nuanced view of these power relationships. To do this, we frame certification systems as sites of constant negotiation and resistance through which Southern actors are able to take up new subject positions in light of their engagement. Drawing on the experiences of two distinct groups of smallholder producers engaged in South-North horticultural trade – ethical/fair trade French bean producers in Kenya and organic/fair trade pineapple producers in Uganda – this paper explores the capacity of smallholder farmers to renegotiate trade relations in ways that represent their interests, and the place that certification systems might play in this. We focus on the cases of 'Kenya-GAP' and 'Ugo-Cert'; two sites where producers and/or their national organizations have engaged in processes of resistance and negotiation with regulatory and accreditation systems in order to have their local interests recognized. We argue that while these case studies reveal a willingness on the part of some regulators and civil society to expand the scope of organic, fair and ethical standards, for structural and cultural reasons, smallholders themselves have often remained removed from these negotiations. Our research also reveals, however, that some alternative spaces for African smallholder resistance exist – demonstrated in a diversity of strategies and activities associated with non-export food systems – but that these are systematically excluded from transnational food network governance.

## Organic, fair and ethical trade for African smallholder farmers

In response to the growing global demand for food produced 'ethically' from Africa, smallholder involvement in export horticultural trade has increased dramatically since the late 1980s. Historically, many of these food producers have been made dependent on export markets, often with negative outcomes – including Southern import dependence, food insecurity, hunger and poverty (McMichael, 2004). Today, they are once again facing pressures to 'upgrade' from traditional forms of agriculture in order to participate in global trade, leading to an intensification and diversification of smallholder farming towards high-value horticultural crops (and increasingly to the production of agro-fuels, see, for example, McMichael, this volume). African smallholders have also had to extend their role in the supply chain, often engaging in cleaning, sizing, trimming, and packing (i.e. value-adding), thus integrating new actors – namely women – as employees in highly gendered and inequitable globalized supply chains (see Barrientos and Dolan, 2006).

As a development strategy, this shift towards export-oriented agriculture in Africa has been implemented with the aim of improving smallholder incomes and livelihood sustainability through increased market access. Certification to governance instruments such as organic, fair and ethical trade is positioned as crucial for inserting African smallholders into export markets in ways that address the power differentials and injustices which have historically characterised globalizing markets. While these schemes emphasize the important role of trade in improving the livelihoods of *all* producers, smallholder farmers[3] have been identified as a particularly difficult group to regulate via 'one size fits all' standards and regulations. Smallholders may or may not belong to farming cooperatives, national labour unions or other collective bodies, and may not be protected by national labour laws (Ethical Trading Initiative, 2005). As such, smallholders are among the most vulnerable actors in global food networks, despite the potential of smallholder farming to solve problems of local food security and environmental degradation.

There is much evidence suggesting that standards may work against the interests of these producers. In Africa, the high costs of accreditation to certifications are often borne by producers, rather than supermarkets and/or exporters; and in circumstances where exporters and/or other stakeholders do cover certification costs, smallholders' certification is tied to these actors, creating new forms of dependency. This creates a barrier for many smallholder producers to participate in these food networks (see Borot de Battisti *et al.*, 2009). Across the literature, the story seems to be the same: while some Southern participants in these emerging standards and accreditation systems have benefited in some labour-related aspects, these are extremely dependent on commodity, country context, employment status, and gender relations (Barrientos and Smith, 2006). More broadly, the export-oriented expansion of organics, fair and ethical trade in the South has been critiqued for centralizing a neoliberal and privatized development agenda, one in which livelihood outcomes in the South – including food security – are tied to individual consumer preferences in the North, and to the 'benevolence' and 'stewardship' of powerful Northern actors (Blowfield and Dolan, 2008; Dolan, 2010). Meanwhile, the food sovereignty movement has called for the re-orientation of agriculture and food production away from export markets towards feeding local communities (Holt-Gimenez *et al.*, 2009).

In the context of these debates, horticultural food production in Africa has clearly emerged as an important site at which to engage with questions of smallholders' participation and power in global food networks. For Marsden (2000, p27), the low levels of participation by Southern smallholders in defining the values embedded in organic, fair and ethical trade represents an 'incipient process of social and political marginalization in current food/power relationships'. This literature, however, presents just one side of the story. Concurrently, certification systems – and the processes through which smallholder producers are incorporated into those systems – also provide sites of ongoing contestation and negotiation of the social and ecological values by which these standards operate. As Friedmann and McNair (2008) have identified, global agri-food relations are being shaped both from above (such as in the case of supermarket ethical trade) and from below (as in grass-roots regional organizations for organic or fair trade). And as export horticulture for smallholders is usually accompanied by subsistence and local market production, the livelihood priorities underpinning these food systems form an integral component of the sustainability needs for smallholders engaged with export horticulture and its certification schemes. Following a brief overview of the research methodology, the remainder of this chapter will illustrate that while the particular versions of sustainability discourse represented by organic, fair and ethical trade have been established as legitimate and integrated into agri-food regulatory systems, Southern producers and producer organizations have demonstrated that it is possible for these concepts to be articulated in diverse – and sometimes contradictory – ways.

## Research methodology

The findings presented in this paper draw from field research conducted by the authors between 2004 and 2009 in Kenya (see Figure 12.1) (Smith) and Uganda (Lyons). Both of these countries have experienced significant growth in export production of organic, fair and ethical trade horticultural products, directly affecting the livelihoods of smallholders or outgrowers who grow the majority of certified fresh horticultural products (with the exception of flowers). In Kenya, French bean production (along with other baby vegetables) goes mostly to the European supermarket sector, and is regulated by a number of voluntary, multi-stakeholder codes for good agricultural practice and ethical labour standards coming out of Europe, such as GLOBALG.A.P and the Ethical Trading Initiative (ETI) (IIED and NRI, 2008; Willer et al., 2008). Of these exports, very little is certified fair trade (Fairtrade Labelling Organizations, 2009), although the recent accreditation of one of Kenya's largest horticultural export companies signifies the potential for fair trade to grow in Kenya, despite the global financial crisis. By contrast, Uganda (see Figure 12.2) has the largest area of land under certified organic production in Africa (an estimated 296,203 ha), the majority of which, similar to fair and ethical trade, is exported to Europe and certified to EU organic standards (Willer et al., 2008). Uganda also produces 18,000 tonnes of fair trade coffee annually, as well as ethically traded vegetables such as green beans, snow peas, courgettes, baby corn, asparagus and squash, while some crops are certified with a number of certification standards, including organics and fair trade (Barrientos and Dolan, 2006).

This makes Kenya and Uganda excellent sites for researching how global certification schemes are being resisted and negotiated by smallholders and their organizations at the national level. Each country also has a number of national-level

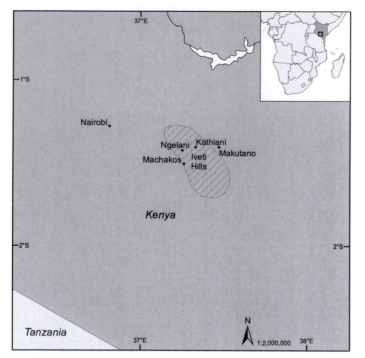

*Figure 12.1*
Map of Kenya
showing
approximate
location of Iveti
Hills

*Figure 12.2*
Map of Uganda
including
Mt. Elgon
National Park
and approxi-
mate location
of Kyaganza
sub-county

industry representative bodies and their own fair or ethical trade associations,[4] which actively lobby for changes to global standards. In Kenya, data collection consisted of participatory action research workshops and in-depth interviews with over 60 smallholders in the Machakos district, as well as representatives of NGOs or national-level industry bodies involved in ethical trade in Kenya in 2007. These farmers were producing ethical trade french beans, courgettes and baby corn for export to the UK under GLOBALG.A.P and the ETI base code (see Blowfield and Dolan, 2008). At the time of data collection, some of these producers were also exploring potential accreditation to fair trade.[5] In Uganda, fieldwork included in-depth interviews with 30 smallholder organic farmers certified by international organic certifiers (including IMO and KRAV) and through both international and domestic (Ugo-Cert) inspectors. Many of these farmers were members of the Katuulo Organic Pineapple Cooperative Society (KOPCS), located in the Kyazanga Sub County, south west of Kampala. Interviews were also conducted with representatives from Amfri Farms (the horticultural export company), as well as NGO representatives in Uganda. Qualitative data were analysed via thematic coding.

In the sections that follow, we focus on two examples where farmer or industry organizations in Kenya and Uganda have been able to renegotiate South-North trade relations, through (1) the process of developing Kenya-GAP as an equivalent voluntary standard to GLOBALG.A.P, and (2) the formation of Ugo-Cert for the provision of local organic certification in Uganda.

## Ethical trade in Kenya: the case of Kenya-GAP

For Kenyan horticultural producers, accreditation to GLOBALG.A.P (previously EurepGAP[6]) and the ETI provide the main entry points into the European market: while voluntary, these are commonly perceived as basic requirements for African suppliers (see Borot de Battisti et al., 2009). EurepGAP emerged in the mid-1990s as a way for supermarkets to blend environmental sustainability and food safety into standards to assure increasingly scrutinising consumers of the credibility of suppliers' good agricultural practices (GAP). The certificate covers the process from farm inputs such as feed or seedlings, through all the farming activities until the product leaves the farm, and includes ethical, social, environmental and labour-related criteria (GlobalGAP, 2006). The Ethical Trading Initiative[7] focuses specifically on labour issues (health and safety, employment-related rights), and complements the GLOBALG.A.P approach to good agricultural practice.

As one example of an ethical trade standard, GLOBALG.A.P has been described as 'a "mega-audit" of European supermarket supply chains' (Campbell and Le Heron, 2007, p132). The private sector body aims to establish one standard for good agricultural practice with different product applications fitting the whole of global agriculture – currently, there are more than 100 independent and accredited GLOBALG.A.P certification bodies in more than 80 countries worldwide; in Kenya, there are 386 certified supplier groups, second only behind South Africa

(IIED and NRI, 2008). In this case study, 600 smallholders were accredited to both GLOBALG.A.P and the ETI base code through the export company. This exporter has also recently gained fair trade accreditation for some of its smallholders.

According to GLOBALG.A.P, their standard serves as a practical manual for good agricultural practice anywhere in the world, based on equal partnership between agricultural producers and retailers. However, the audit process includes complex requirements for record-keeping, high amounts of technical knowledge, and adherence to advanced labour and welfare laws, all based on a European style of agricultural management (Campbell and Le Heron, 2007). A large body of research reveals that while GLOBALG.A.P has brought some improvements, African smallholder farmers remain disadvantaged by the costs of compliance, chain management pressures and threshold effects. Similar criticisms have been made of the ETI and fair trade. As a result, many smallholders have shifted to markets which do not require GLOBALG.A.P compliance. In Kenya, smallholder involvement in export horticulture has reduced by 60 per cent since 2002 due to the costs associated with ethical accreditation (Borot de Battisti, *et al.*, 2009, p41; IIED and NRI, 2008, p12). As such, GLOBALG.A.P was viewed not only as a major barrier to trade for Kenyan smallholders, but also as overly eurocentric, and as having little benefit for the livelihoods of smallholders themselves. Similar tensions clearly surround the ETI smallholder guidelines, considering that discussions of impacts for African smallholders were completely absent from the most recent ETI conference in 2008.

According to a representative of the Fresh Produce Exporters Association of Kenya (FPEAK), the first version of EurepGAP (and thus later GLOBALG.A.P) 'had nothing to do with the African setting'. This has prompted the recent growth of an 'equivalence movement' in global standards – whereby national standards are benchmarked against the GLOBALG.A.P standard. In Kenya, this has taken the form of Kenya-GAP.[8] For FPEAK, Kenya-GAP represents an effort to provide a universally accepted standard based on the principles underpinning GLOBALG.A.P, but which is customized to meet the needs and priorities of local farming conditions, including smallholders. However, while FPEAK was successful in gaining equivalence on some issues, smallholders themselves remained removed from the decision-making process. It emerged that this was largely due to smallholders' under-representation in the formal channels of consultation, at the same time as organizations better able to represent smallholders' interests were excluded. This has meant that smallholders are largely spoken *for*, rather than speaking for themselves.

### Negotiating sustainability values

Following an outcry from non-European horticultural bodies and lobby groups, FPEAK (as members of GLOBALG.A.P) played a significant role in the revisions of the GLOBALG.A.P code, which in 2006 became Kenya-GAP. This outcry occurred over both the content of the original standard and the lack of participation of Southern actors in its development, reflecting scepticism around the eurocentrism of the standard. FPEAK was aware that many of the costs of accreditation to GLOBALG.A.P were restrictive for smallholders in the french bean trade. For example, GLOBALG.A.P's elaborate requirements for the safe handling and containment of chemicals – requiring a fire resistant chemical store specially built to avoid spillages – were financially out-of-reach for small-holders. This complex structure far exceeded the value of smallholders' crops, leading to conflict over the meaning of sustainability. It was also culturally inappropriate, as the chairperson of the Kenyan EurepGAP National Technical Committee explained:

> some of the farmers that you are asking to have that elaborate structure ... their own houses are not worth that, as much as that structure. So why would you want me to construct such a thing while my social status does not warrant me. Because the little income the person is getting is paying medicine, is paying school fees, whatever. For the family. So it's not going to be sustainable.

FPEAK was successful in convincing GLOBALG.A.P that the standard needed to deal primarily in the basic principles of good agricultural practices. According to FPEAK, the principles of good agricultural practices are very easy to understand:

> One, why do you want to manage the spillage? Because of the environment ... Two, why do you want people to be protected? Because of health and safety of the people. Why do you want the place to be lockable? Accessibility.

Convincing GLOBALG.A.P of the technical and scientific legitimacy of these ideas was crucial for supporting alternative visions of good agricultural practices that are fundamentally about sustainable agriculture. Thus, based on values around protecting the health and safety of both environment and people, Kenya-GAP now stipulates that a small metal box is sufficient to replace expensive storage rooms, and is a more economically- and culturally-acceptable means of preventing chemical spills (Borot de Battisti et al., 2009). This is because smallholders' 'net income is not going to be eaten by a structure' (FPEAK interview, 2007).

Kenyan smallholders are now also able to prove the traceability of products as required by GLOBALG.A.P (as well as the EU directive on traceability) through improved record-keeping procedures that fit with local contexts. Kenya-GAP simplifies this process so that good practice in this realm refers simply to noting down any farm activity implemented in a small farm diary. Similarly, the ETI smallholder guidelines recommend actions that individual smallholders and coop-eratives can take to simplify record keeping, although the extent to which small-holders have access to records specifically related to contracts or wages is limited. According to FPEAK, it is not important for smallholders to be bombarded with technical regulatory information, because 'once you put it that way, then the farmers own the process, they know it is also helping them'. This discourse of farmer participation in 'owning the process' is strong throughout FPEAK's discussion of the Kenya-GAP success story. But while FPEAK itself was able to gain recognition for equivalence based on the values of sustainability rather than on the technical instruments with which to achieve these, the question remains: what role did smallholder farmers themselves play in these negotiations?

### *Smallholder participation*

According to FPEAK, the Kenya-GAP revisions 'have to involve everybody, all the dissenting voices, including NGOs and governments and other people, practitioners'. When asked about the extent of smallholder participation in the Kenya-GAP lobbying process, it becomes apparent that there were both structural and cultural constraints to smallholders' active participation. In FPEAK's view, while the technical committee did a number of 'mock audits' of farms that inevitably employed smallholders, they admit that farmers from very small farms did not participate to a great extent. They, apparently, participated by default, in that the committee was 'a representative committee from technical people working with various farmers across the globe'. FPEAK also claimed that the involvement of women was strong, simply 'because some of us have worked with women for quite some time'.

While personal histories of involvement on-farm with smallholders informed how FPEAK approached the issue of smallholder participation, in practice, the selection of farms for consultation followed the pattern by which smallholders are selected by horticultural export companies. For example, Kenya's largest (and arguably, most respected) ethical trading horticultural export company officially requires smallholders with five or more acres of land in order to create the economies of scale necessary to ensure sufficient continuity and quantity of produce to offset the costs of accreditation for multiple small farmers. This potentially locks out very small-scale outgrowers. As all the smallholders in this research owned less than five acres, they would have undoubtedly been among those excluded from the discussions. Indeed, only one group out of five interviewed were actually aware of a recent audit or visit being conducted on their farms. Thus, the structure of exporter-smallholder relations meant that while the Kenya-GAP negotiations attempted to represent every category of grower, in reality the majority of contact was via medium-sized exporters.

Although FPEAK wanted to generate a Kenya-GAP which was bottom-up and not foreign, most Kenya-GAP changes appear to have come from FPEAK's technical experience in documenting what smallholders are doing on their farms. In the chemical storage box example discussed earlier, the idea for equivalence came from observations that farmers were already successfully storing chemicals safely in metal boxes. This is in direct contrast to participatory auditing that Auret and Barrientos (2006) advocate for dealing with smallholders in Africa. As they explain, auditing which focuses on formal management 'snapshots' is useful for picking up visible issues such as health and safety, but less useful for finding out about more complex ethical issues such as freedom, discrimination or culturally-based inequalities (Auret and Barrientos, 2006). The negotiations also did not necessarily open up spaces for smallholders themselves to comment, largely because FPEAK formally involved producer cooperatives but did not address existing inequalities within these.

### *Producer cooperatives and their alternatives*

Ethical trade regulations are enacted at the local level through interactions between the exporting company – whose responsibility it is to monitor and enforce standards – and smallholder producer cooperatives. These groups, used in the majority of green bean export production, have been described as playing 'a major part in the success that has been experienced in [Kenya's] horticultural industry' (Borot de Battisti *et al.*, 2009, p27) by providing opportunities for farmers to pool produce and resources, facilitate market access and increase knowledge of market requirements such as GLOBALG.A.P and the ETI. In brief, smallholder farmer cooperatives were set up by exporting companies for the purposes of organizing and informing small groups of farmers in each area, and thus formed the basis for the Kenya-GAP technical committees.

The issue, however, is that cooperative membership often privileges land owners. The most marginalized smallholders – i.e. farm workers and women (who contribute 80 per cent of the labour) – were found to be excluded from most technical training, information sessions and decision-making opportunities associated with farmer cooperatives. Even where these farmers participate, for example as employees involved in record keeping or in sorting and grading produce, they are not often formal members. For women in particular, their husbands often attend meetings instead. FPEAK suggested that this problem 'had to do with the cultural setting more than anything else'.

As a result of poor representation in producer cooperatives, smallholders' knowledge about aspects of the food system beyond the production level was extremely low. Women smallholders especially had little understanding of the global market for french beans; some were aware that the beans were exported, but did not know the final destination, although they realized that produce was destined for foreign consumers due to the fact that french beans were not part of the local diet. They had trouble selling surplus on the local market for a decent price, and did not know how to prepare the food for family consumption. This meant that high quality 'ethical' surplus or reject vegetables were commonly used as animal feed, and farmers lost valuable income. This has important implications for the role of standards such as GLOBALG.A.P and Kenya-GAP to contribute to improving the livelihoods and food security of smallholder farmers at the national and local level.

Smallholders were also unaware of any global regulatory standards, beyond immediate health regulations. When asked about specific parts of GLOBALG.A.P or the ETI code for example – such as collective bargaining, freedom of association, fair wages and so on – many responded with, 'you know, I don't know many rules'. Smallholders were not familiar with the process of auditing or accreditation, despite noting that sometimes representatives from the exporting company can be seen walking around their shambas (farms). In the case of one farm that had recently experienced what was perhaps an audit, even the farm manager had very little knowledge of what this entailed:

> I think there was something like that. Because there was time when there was some white people just like you came and wanted to see the standard of how we produce, how we grade. So they wanted to see how we do it ... Yes, they were coming with videos.

Kenya-GAP negotiations did not open up spaces for smallholders themselves to comment, largely because FPEAK formally involved producer cooperatives but did not address existing inequalities within these. This problem could have been addressed by incorporating other groups where marginalized smallholders (rather than male land owners) are more active and empowered, into the Kenya-GAP negotiation process. Examples of this type of group are *maendeleo* or *harambee* groups – self-help-style groups – which emerged as a much more effective strategy for women smallholders especially to assert their agency at the local level. In these groups, smallholders can challenge cultural norms that otherwise result in their exclusion from decision-making in producer cooperatives; indeed they were so important that smallholders described these groups more like 'actors' in the food system than producer cooperatives, because their activities go beyond market access to include providing labour, financial assistance, information and social support. The implications of excluding such groups from the Kenya-GAP process was evident in that, while smallholders themselves chose to discuss a wide range of health and safety, environmental and social concerns, none of these issues emerged as key areas of concern to FPEAK in regards to Kenya-GAP. FPEAK committees were interested in localized solutions to problems of affordability of implementation that could be quantified in the language of audits, rather than ones that required more complex negotiation of cultural and structural constraints on smallholders.

This chapter now turns to consider similar challenges affecting organic smallholder producers in Uganda, suggesting that while equivalence movements in standards involve some improvements for smallholders, they are restricted to governance channels from which smallholders remain marginalized.

## Organic agriculture in Uganda: the case of Ugo-Cert

The first organic export projects from Uganda[9] began in the early 1990s, and included organic solar-dried tropical fruit, chillies, ginger, beans and okra, and an organic cotton export initiative. Since then, organic exporting has expanded to include a range of other products, including vanilla, bark cloth, cocoa, Arabica and Robusta coffee and sesame. In almost all cases, farmers are certified organic via a group certification process. Many export initiatives in Uganda have received financial assistance from the Export Promotion of Organic Products from Africa (EPOPA) – with funding via the Swedish International Development Cooperation Agency (SIDA) – and organic certification is provided by a number of international organic certifiers, including IMO, EcoCert, BCS and KRAV. In addition to organic certification, many products are also certified with other social and environmental standards, including Utz Certified, Fair Trade and GLOBALG.A.P. Compliance with organic standards is a requirement for access to the international market for organic produce and, in this context, and similar to fair and ethical trade, debates have emerged related to equivalence and alternative certification systems – including local certification systems.

### *Localizing Organic Auditing*

Compliance with international organic standards has been widely critiqued for its high cost to farmers, especially smallholders; its reliance on culturally inappropriate audit requirements (including detailed written record keeping); and the limited representation of smallholder farmers and other actors from the South in processes related to standards setting (Lyons *et al.*, 2012 forthcoming). According to Dolan (2010, p41), these circumstances have left smallholders and farming organizations in the South with 'constrained possibilities' in international trade negotiations.

In large part as a response to these concerns, a national peak body for the Ugandan organic movement was established in 1998. The National Organic Agriculture Movement of Uganda (NOGAMU) attempts to provide a collective regional and national voice for the organic sector, and in this role has played a part in the formation of the domestic certification organization Ugo-Cert, and the domestic organic standard. As part of its advocacy, NOGAMU – and the organizations, commercial interests and farmers it represents – along with IFOAM (International Federation of Organic Agriculture Movements) and other international and national organizations, have transformed audit arrangements related to organic certification, including the introduction of locally specific audit requirements.

The audit model characterizing organic certification in the North has been widely identified as ill-suited for use in the South (see, for example, Mutersbaugh, 2004; Dolan, 2010). There have been strong calls in Uganda (and elsewhere) to restructure the organic audit culture – characterized by individual-based certification and high levels of written record keeping – to better reflect the diverse cultural and economic circumstances of smallholder farmers: ' ... early on, they were trying to use audit systems developed for EU large farmers rather than smallholders, so we ended up trying to develop something that was a bit more appropriate' (Agricultural consultant).

On the basis of this critique, smallholder group certification was devised. This alternative audit arrangement is based on the organization of smallholder farmers into groups, and with an organic certificate awarded to the group, rather than individual farmers. Management of the group occurs via an Internal Control System

(ICS): an internal quality control document that stipulates requirements related to growing methods, post-harvest handling, record keeping and other activities. An agricultural consultant described the ICS model as working in the following way:

> The certifying body will take a sample of documents from the ICS, cross-check those with the actual farmers, make sure the farmers actually exist, make sure the records we have on the farmers are actually similar to what the farm actually looks like … and then the certifier basically says, "yes, I think this group can manage its own control", or "no, I don't think the group can manage". … The traditional European system was that the inspector works directly with a farm and inspects a whole farm. Now obviously you've got 6000 small farmers, you can't have a *muzungu* (white person) wandering around for the whole year.

This group audit model has provided a mechanism for smallholders to obtain organic certification, circumstances that have been associated with social and economic benefits for some smallholders. At the same time, smallholder group certification has also delivered clear benefits for food processors and food export companies by assisting to secure increased access to certified organic produce. At least 18 export companies in Uganda have received financial support from EPOPA to assist in the costs of organic certification. At the same time, however, the organic smallholder certification scheme has further institutionalized standards compliance as a precursor for market entry. And while smallholder group certification has enabled some smallholder farmers to obtain organic certification, many have also been excluded. Farmers in the Katuulo Organic Pineapple Cooperative, for example, stated there were many smallholders in the district who wished to join the Katuulo Organic Pineapple Cooperative – however, the Cooperative were generally unwilling to expand their membership given they already produced a surplus of organic pineapples; circumstances which left many farmers trying to sell their organic produce on the local market.

On the one hand, the formation of the smallholder group certification system and ICS audit arrangement better reflects the lived reality of farmers in Uganda, the majority of which are smallholder farmers. On the other hand, many smallholder farmers also remain unaware and/or unable to participate in group certification initiatives. At the same time, ICS arrangements wed smallholder farmers to the export company they supply, circumstances that restrict them from selling their produce elsewhere.

### *Localizing Organic Inspection*

In addition to the introduction of audit arrangements specific to smallholder farmers, the Ugandan organic sector – in collaboration with international organizations including IFOAM and EPOPA – has also succeeded in localizing organic inspection arrangements. This represents a shift away from the prior reliance on inspectors from international organic certification organizations, including IMO and KRAV. In 2004, Ugo-Cert was successful in advocating for local inspectors to undertake organic inspection on behalf of international certifiers. Representatives from Ugo-Cert and EPOPA discussed a range of benefits for smallholder farmers associated with local inspection, including a reduction in the cost of inspection; year-round availability of organic inspectors; and the increased likelihood that inspection processes and inspectors will be cognizant of local social, cultural and other contexts. A representative from Ugo-Cert explained that export companies face pressure to utilize local inspectors, and funding from EPOPA was increasingly contingent upon it as well:

> Many operators here in Uganda are actually putting pressure on their certifiers to use the local inspectors, so that the costs can be reduced. So, many of the operators are refusing to meet the costs of flying in an inspector from the UK, or Germany.

The uptake of local inspection – an outcome of lobbying from Ugo-Cert, NOGAMU and a number of farming organizations – has delivered a number of benefits to export companies, including reducing the cost of organic inspection. At the same time, the availability of local inspection provides year-round availability of inspectors for the provision of organic certification, enabling export companies to enter the organic market more quickly.

Yet while each of these aspects has provided economic benefits to export companies, the extent to which smallholders have actually driven such changes – thereby ensuring emerging certification arrangements are sensitive to local cultural needs and aspirations – is less clear. In contrast, much evidence exists to intimate that smallholders' interests were excluded from transformations occurring in organic certification arrangements. Participant observation at an audit training day with international organic inspectors and local field officers (who were being trained on how to inspect farms to ensure compliance with the ICS), for example, demonstrated the extent to which Northern actors continue to shape organic certification processes *for* Southern smallholders, rather than in collaboration *with* them. This was especially evident in terms of attempts to ensure the gender sensitivity of ICS compliance mechanisms. It was in this context that an international organic inspector asked the only woman field officer present if she, or other women she knew, had concerns with the ICS – with specific reference to its sensitivity to women's needs and interests. The woman field officer remained silent, while many men present laughed (presumably at the inappropriateness of this question being asked in a public space, and with men in attendance). While there are many gendered dimensions related to engagement in organic farming that this woman field officer might have discussed (including land tenure, division of labour, access to income, etc.), she could not readily discuss such issues in this public context. To make matters worse, women were significantly underrepresented at this meeting – with only one Ugandan woman present out of a total of 12 local representatives – circumstances that further constrained the extent to which this woman might have discussed gender-sensitive issues. On the one hand, this demonstrates the limits of Northern organic inspectors' understandings of the extent to which gender (amongst other culturally specific contexts) shapes the organization of social life, and distribution of, and access to, resources. On the other, it also points to the limited extent to which local-level actors, including smallholders themselves, have been able to shape emerging processes related to organic certification in ways that better reflect local-level interests and experiences.

Similar frustrations regarding the limits to which local-level interests were recognized and included in the formation of local audit and inspection arrangements related to the actual content of organic standards. One of the primary objectives of the domestic certifier, Ugo-Cert, is to 'actively facilitate the development of realistic and acceptable environmentally sustainable production standards in Uganda and the rest of the world' (Ugo-Cert, 2010). Despite this mission statement, and in contrast to Kenya-GAP's success in achieving recognition related to some locally equivalent standards, interviews with representatives from Ugo-Cert indicated they had, in reality, achieved little success in negotiating equivalence between national, regional and international standards. A representative from Uganda's organic certification company Ugo-Cert expressed his frustration in this regard: 'We have no bargaining power, we have absolutely no say in international negotiations'. He detailed a number of examples where Uganda had little bargaining power in international negotiations, including a proposal for recognition of equivalence for practices related to cattle grazing. In this example, even after years invested in a negotiation process, IFOAM and international certifiers rejected Ugo-Cert's proposal.

### Smallholder farmer exclusion

The discussion presented here demonstrates strategies to incorporate the specific needs and interests of local stakeholder investments within organic certification arrangements via the construction of locally-specific auditing and inspection arrangements. The development of these strategies, however, has largely occurred without consultation with, or inclusion of, those immediately affected by them: smallholder farmers. As a result, group smallholder certification schemes are not always sensitive to smallholder needs, especially those of women smallholders. And while there are now internationally recognized local inspection arrangements, local inspectors may be constrained by international frameworks related to the ICS and its implementation. Similarly, local inspectors are frequently required to ensure smallholders' compliance to internationally-derived standards, given the limited success from the Ugandan organic movement to shape the content of organic standards.

The issues discussed here point to the extent to which smallholders and other local actors have been left out of the design of organic auditing processes, despite these processes specifically emphasizing local-level relevance of organic certification systems. In addition, there was evidence to demonstrate smallholders were excluded throughout the implementation process. For example, smallholder farmers who were members of the KOPCS stated they knew little about the details of the standards they were compliant with, or the requirements of the markets for which their produce was destined.

Demonstrating this disconnect between the content of organic standards and smallholder farmers' practices and knowledge, many smallholders explained they did not know if it was possible to grow certain vegetables organically, and therefore avoided growing these for household consumption in case this jeopardized the organic certification status of their cash crop. Some smallholders also stated they had refrained from experimenting with new vegetable crops out of fear they would lose their certified organic status. While smallholders who were members of the KOPCS had received information and training from Amfri Farms (the buyer) related to growing pineapples organically, this information had not extended to organic growing practices related to non-cash crops (including tomatoes, red onions and pumpkin). As a result, smallholder farmers had come to rely increasingly on the local market for the purchase of some of their basic vegetable needs.

Smallholder farmers also expressed frustration at the separation between themselves and the consumers of their produce. Smallholders lamented that consumer preferences and market requirements seemed to 'fall from nowhere', that they were constantly changing, leaving them sometimes unable to comply. One smallholder held up wrinkled and tattered black and white photocopied photographs of fruit, and explained this was the market information they received from Amfri Farms as advice on quality and size. The photographs themselves displayed bananas with blemishes and marks, and pineapples of various sizes. The farmer lamented:

> How can we know about your markets? How can we know what you want when all we have to go by are such images? I want to go to your country, to learn more about what you *muzungu* (white people) want.

Smallholder farmers were frustrated by the limited attempts to involve them, the outcomes of which frequently resulted in adverse economic impacts, for example through loss of market access. Smallholders also stated that poor engagement and exclusion from access to information maintained their dependence on both the certifier and the export company, with many expressing frustration about knowing little about where their produce was destined for after leaving their farm.

## Discussion and conclusion

In both Kenya and Uganda, smallholder farmers are working within the context of international standards for organic, fair and/or ethical trade, and are part of a growing equivalence movement whereby local and global-level actors are able to negotiate changes to global standards to better reflect African contexts. Through exploring smallholders' contributions to Kenya-GAP and Ugo-Cert negotiations, this chapter has demonstrated a number of similarities and differences between the Kenyan and Ugandan experiences.

In Kenya, there has been some success in altering ethical standards to better reflect local specificities. Namely, the national industry body for horticulture, FPEAK, was successful in convincing GLOBALG.A.P of the need to refocus on underlying values of sustainability such as good environmental practices; health and safety; affordability (as exemplified by the acceptance of chemical storage facilities that better suit smallholders' budgets); and simplified record keeping practices for proving compliance to newly benchmarked Kenya-GAP. The same cannot be said in Uganda, however, with few practical outcomes of equivalence of direct benefit to producers. Uganda has, however, had greater success in altering certification and audit processes more generally, through the introduction of a domestic organic standard, local inspections and group small-holder certification through Ugo-Cert. This is promising, in that the group certifi-cation model is set to spread from Uganda, and beyond organics to other quality standards (GLOBALG.A.P, Fair Trade, etc.). Similarly, the development of a joint inspection protocol in Uganda has provided a precedent for the establishment of protocols elsewhere, including the recently drafted Regional Organic Agriculture Standard in East Africa, as well as standards for other African nations. All of these recent changes have opened new opportunities (and challenges) for smallholders to negotiate and resist definitions of ethics, quality and sustainability in South-North horticultural trade.

While FPEAK and Ugo-Cert were positive about the outcomes of changes to the content and implementation of ethical or organic standards for smallholder farmers, our findings reveal that smallholders are largely *spoken for*, rather than speaking for themselves. In Kenya, this resulted from structural barriers to smallholders' participation in the channels through which national sectoral bodies such as FPEAK have used to 'consult' farmers. The equivalence process failed to involve the groups in which smallholders themselves are most active, and instead relied on engagement with a limited selection of producer coopera-tives. Unfortunately, equal and informed participation in producer cooperatives is not ensured for all smallholders, thus resulting in a quite narrow and technical set of changes in Kenya-GAP. Because exporters provide the 'link' between smallholders, industry representatives such as FPEAK and global ethical trade governance institutions, the structural barriers within cooperatives are repeated throughout the supply chain, with adverse implications for smallholders' knowl-edge, participation and influence. By contrast, Ugo-Cert was better able to introduce cultural sensitivity into the organic audit culture in Uganda, through local inspections. However, these inspection processes remain constrained by Northern-imposed structures for local-level engagement, the outcomes of which can result in processes that are insensitive to cultural – including gender – specificities.

In Uganda, although group certification has the potential to empower producer cooperatives vis-à-vis global standard bodies, smallholders argued that this had not reduced the practical problems that dependence on exporters had for smallholders in those cooperatives. Lack of information, lack of market access and oversupply of produce were problems that continued to affect the Katuulo producer cooperative despite the changes to local auditing and inspection processes. These problems were likewise experienced by Kenyan smallholders, suggesting that both Kenya-GAP and Ugo-Cert represent localized solutions to problems that can be quantified in the language of audits, rather than ones that require more complex negotiation of cultural and structural constraints on smallholders. This limits the broader livelihood benefits of smallholder engagement in global trade, and restricts the flow of benefits into local food systems as well. As suggested by Dolan (2005, p424), these kinds of experiences raise important questions about 'the types of development they actually engender and whose interests these outcomes serve'.

At first glance, these findings tend to support criticisms that ethical, fair and organic standards reproduce market-based inequalities through audit technologies that advance global neoliberalism. Although standards are often construed as governance instruments non-specific to geography, culture and commodity, Dolan (2008) argues that the audit process actually authorizes and naturalizes categories (which measure social and environmental values) that have often been developed without the input of producers in the South, thus reproducing Northern ethical values and empowering Northern actors in the process. Neither FPEAK nor NOGAMU had been able to address these systemic inequalities at the local level. The smallholder exclusion reported in this chapter has demonstrated, to some extent, exactly these inequities. Mutersbaugh (2004) characterizes the exclusion of smallholder farmers in these negotiations as a form of eco-colonialism; referring to the imposition of Northern retailer and consumer interests upon the livelihoods of Southern farmers. As such, while alternative food networks – and the certification systems that underpin them – are frequently discursively framed around notions of 'empowerment', 'participation' and 'partnership' (see also Dolan, 2010), this shadows issues of lack of power, exclusion and inequality experienced by smallholder farmers.

However, more positively, our discussions with smallholders have also shown them to be active agents of change. There is a great diversity of localized knowledge, experiences, impacts and challenges associated with smallholders' participation in, or exclusion from, ethical trade in the South. Our findings reinforce how closely related institutional participation by Southern food producers is to the depoliticization of organic, fair and ethical trade agendas that has resulted from their standardization through neoliberal technologies such as audits, standards and accreditation systems (see Dolan, 2010). The challenge remains for equivalence processes to go beyond these structures in order to draw from sites where smallholders are actively exerting their own agency, knowledge and priorities for the more environmentally sustainable, fair and ethical management of their food systems. It is here that lessons for shifting certification systems in order to better meet the needs of smallholders exist, and it is through these activities that rethinking South/North relations will be possible.

In conclusion, while institutional structures and cultural barriers can certainly restrict the participation of Southern producers and organizations in regulatory decision-making processes, in other instances – including our case studies – smallholders' activities and interests have successfully informed changes in how these transnational food networks are governed, with some benefits for smallholders. These attempts to reshape globally-defined food production and trade conventions to better incorporate the needs of Southern producers, NGOs and other actors also reveal a willingness and flexibility on the part of some regulators, and civil society, to expand the scope of organic, fair and ethical standards. African smallholders should be seen as political subjects and actors in local-global economies in which organic, fair and ethical trade are partial components. In Africa, the full extent to which local institutions, industry bodies and smallholder collectives are able to shift the content of organic, fair and ethical standards to better address the structural and cultural inequalities underpinning the interconnection between trade and livelihoods, and the impacts of such changes for local – and indeed global – food security remains to be seen.

## Notes

1 Definitions of smallholders vary immensely in the literature and across horticultural industries. Often, the terms 'smallholders' and 'outgrowers' are used interchangeably. In this study, smallholders are defined as producers who grow relatively small volumes of produce on relatively small plots of family-owned and -managed land, often combining production of an export commodity with other livelihood activities.

2 The terms 'organic farming' or 'organic agriculture' are frequently utilized interchangeably with other concepts, including 'chemical-free', 'de-facto organic' or 'traditional farming practices'. In this chapter, we utilize the term 'organic agriculture' to denote those farming systems that actively avoid the use of synthetically-derived agricultural chemicals and that utilize natural farming practices to build up the capacity of the soil. This may refer to both certified and non-certified organics (see Parrott and Marsden, 2002).

3 Smallholders may: be men or women; depend on family labour or may hire workers; combine high value horticultural crops with the production of other cash crops for sale on the local market and for subsistence; be generally less well-resourced than commercial-scale farmers; and are usually considered as part of the informal economy.

4 In Kenya, these include: the Fresh Produce Exporters Association of Kenya (FPEAK), Kenyan Flower Council, Horticulture Ethical Business Initiative (HEBI), and the Cooperation for Fair Trade in Africa (COFTA). In Uganda, these include: the National Organic Agriculture Movement of Uganda (NOGAMU), Export Promotion of Organic Products from Africa (EPOPA) and Ugo-Cert.

5 By April 2010, this exporter had gained fair trade certification for some of its vegetables due to changes in fair trade rules with regard to sourcing from outgrowers, although it is not likely that participants in this study are among those now growing for fair trade markets.

6   EurepGAP has been discussed in-depth elsewhere, and will not be repeated here. Please see Campbell and Le Heron (2007) for more information.
7   While the ETI (and fair trade, for that matter) have been written about extensively elsewhere (Barrientos and Dolan, 2006), they none-the-less represent an important part of the regulatory context in which the smallholders in this case study operate.
8   Similar processes have resulted in Chile-GAP, Mexico-GAP and China-GAP.
9   The expansion of organic agriculture in Uganda occurred at the end of the Amin and Obote regimes in the mid 1980s. Support for the expansion of organic agriculture was part of a strategy to re-build local food security, which was jeopardized by decades of political instability, conflict and violence. By the early 1990s however, organic agriculture in Uganda — and more broadly in Africa — had shifted from an initial emphasis on local food security to the production of organic cash crops for export markets (Parrott and Marsden, 2002). This export focus has been supported by international development agencies via the provision of various supports to food export companies, including: finance related to the costs of conversion and certification, market access information, infrastructure support and the provision of other resources. In Uganda, the Swedish International Development Corporation Agency (SIDA) funded the Export Promotion of Organic Products from Africa (EPOPA) between 1995 and 2008, supporting (amongst other aspects) the establishment of organic supply chains.

# References

Auret, D., and Barrientos, S. (2006) 'Participatory Social Auditing: Developing a Worker-focused Approach', in S. Barrientos and C. Dolan (eds) *Ethical Sourcing in the Global Food System*, Earthscan, London, pp129–148

Barrientos, S., and Dolan, C. (eds) (2006) *Ethical Sourcing in the Global Food System*, Earthscan, London

Barrientos, S., and Smith, S. (2006) *ETI Labour Code of Practice: Do workers really benefit?*, Institute of Development Studies, University of Sussex, UK

Blowfield, M., and Dolan, C. (2008) 'Stewards of Virtue? The Ethical Dilemma of CSR in African Agriculture', *Development and Change,* vol 39, no 1, pp1–23

Borot de Battisti, A., MacGregor, J., and Graffham, A. (eds) (2009) *Standard Bearers: Horticultural exports and private standards in Africa*, International Institute for Environment and Development, London

Campbell, H., and Le Heron, R. (2007) 'Supermarkets, Producers and Audit Technologies: The Constitutive Micro-politics of Food, Legitimacy and Governance' in D. Burch and G. Lawrence (eds) *Supermarkets and Agri-food Supply Chains: Transformations in the Production and Consumption of Foods,* Edward Elgar, Cheltenham, MA, pp131–153

Dolan, C. (2005) 'Benevolent Intent? The development encounter in Kenya's horticulture industry', *Journal of Asian and African Studies*, vol 40, no 6, pp411–437

Dolan, C. (2008) 'Arbitrating Risk through Moral Values: The case of Kenyan fair trade', *Economic Anthropology,* vol 28, pp271–298

Dolan, C. (2010) 'Virtual moralities: The mainstreaming of Fairtrade in Kenyan tea fields', *Geoforum,* vol 41, no 1, pp33–43

Ethical Trading Initiative (2005), 'Smallholder Guidelines', www.ethicaltrading.org

Fairtrade Labelling Organizations International (2009) *Fair Trade Leading the Way: Annual report 2008-09*, Fairtrade Labelling Organizations International, Bonn, Germany-

Friedmann, H., and McNair, A. (2008) 'Whose Rules Rule? Contested projects to certify "local food products for distant consumers" ', *Journal of Agrarian Change*, vol 8, nos 2 and 3, pp408–434

GlobalGAP (2006) *Generic Manual on Quality Management System for Smallholder Horticultural Farmer Groups in Kenya for Certificiation to EurepGAP Option 2,* Agribusiness & Allied Kenya Ltd, Standards & Solutions Consulting Ltd and ICIPE (Nairobi - Kenya), Nairobi, Kenya

Holt-Gimenez, E., Patel, R. and Shattuck, A. (2009) *Food rebellions: Crisis and the hunger for justice,* Food First Books, Fahumu Books and Grassroots International, Oxford, UK and Oakland, CA

IIED and NRI (2008) *Costs and benefits of EurepGAP compliance for African small-holders: A synthesis of surveys in three countries; Number 13*, IIED and NRI, London

Lyons, K., Palanippan, G. and Lockie, S. (2012, forthcoming) 'Organic Agriculture Governance in the Global South: New Opportunities for Participation in Agricultural Development and Livelihood Outcomes', in N. Halberg and A. Muller (eds) *Organic Agriculture for Sustainable Livelihoods*, Earthscan, London.

McMichael, P. (2004) *Development and Social Change: A global perspective* (4th ed.), Pine Forge Press, Thousand Oaks, CA

McMichael, P. (2010) 'Agrofuels in the Food Regime', *Journal of Peasant Studies*, vol 37, no. 4, pp609–629

Marsden, T. (2000) 'Food Matters and the Matter of Food: Towards a New Food Governance?' *Sociologia Ruralis,* vol 40, no 1, pp20–29

Mutersbaugh, T. (2004) 'Serve and Certify: Paradoxes of Service Work in Organic-Coffee Certification', *Environment and Planning D-Society and Space,* vol 22, no 4, pp533–552

Parrott, N., and Marsden, T. (2002) *The Real Green Revolution: Organic and agroeco-logical farming in the South*, Greenpeace Environmental Trust, London

Raynolds, L. (2002) 'Consumer/Producer Links in Fair Trade Coffee Networks', *Sociologia Ruralis*, vol 42, no 4, pp404–424

Ugo-Cert (2010) 'Mission Statement', www.ugocert.org

Willer, H., Yussefi-Menzler, M., and Sorensen, N. (eds) (2008) *The World of Organic Agriculture: Statistics and emerging trends 2008*: IFOAM, Bonn, Germany and Research Institute of Organic Agriculture, Frick, Switzerland

# 13

# FOOD FOR THOUGHT? LINKING URBAN AGRICULTURE AND LOCAL FOOD PRODUCTION FOR FOOD SECURITY AND DEVELOPMENT IN THE SOUTH PACIFIC

*Alec Thornton*

Throughout the developed and developing world, millions of people engage in urban and peri-urban agriculture (UPA) for a variety of reasons and using different production and distribution systems. In differentiating 'urban' from 'rural' production systems, perhaps it is helpful to ask the question 'what is agriculture'? According to the Oxford Dictionary, 'agriculture' is *the science or practice of farming, including cultivation of the soil, for the growing of crops and the rearing of animals to provide food, wool and other products*. Agriculture taking place in urban settings will have characteristics that distinguish it from rural types of cultivation. Although both types of production systems share certain biophysical and natural determinants, an urban classification includes additional attributes, such as unique productivity constraints (issues of legality, land tenure, pollutants), spatial form (vacant lots, rooftops) and practical function (husbandry, crops, informal source of income and employment).

When observing urban agriculture as a distinct typology, there appears to be significant differentiation in its practice, reflecting varying levels of economic and social development. In the global North it often takes the form of a social movement for sustainable communities, where organic growers, 'foodies' and 'locavores' (those who eat locally-produced food) form social networks founded on a shared ethos of nature and community holism. These networks can evolve when receiving formal institutional support, becoming integrated into local town planning as a 'transition town' movement for sustainable urban development. In the developing South, food security, nutrition and income generation are key motivations for the practice. However, in both areas, it largely exists as an informal activity that is too often discouraged by city planners and health officials who view the practice as a rural activity, misplaced in urban areas, and harbinger of malaria, disease and other public hazards. The persistence of global hunger and malnutrition in the rapidly urbanizing global south could see UPA emerging as a formal initiative.

This chapter will discuss the potential for urban agriculture to emerge as a formal policy for local food production in one of the more geographically and economically isolated regions of the developing world – the small-island developing states of the South Pacific (see Figure 13.1). The following sections will provide an overview of urban agriculture, including conditions where it tends to thrive, as a response to minimize risk, urban poverty and food insecurity. The role of institutions in facilitating UPA and local food production in the South Pacific is explored in case studies of Samoa and Fiji, where the impacts of global-change[1] (O'Brien and Leichenko 2003) are generating differing responses. Finally, the potential of UPA as a formal policy approach to ensure food security and sustainable urban development is considered.

## What is urban agriculture?

Urban agriculture (UA) can be understood as the cultivation, processing, marketing and distribution of food, forestry and horticultural products that occur in built-up 'intra-urban' areas. The broader term, urban and peri-urban agriculture (UPA), is typically used to describe all urban food production systems, both in the built up areas and along 'peri-urban' zones or 'fringes' (also referred to as 'green-belts') of cities and towns on public, private and communal or customary land. In most cases, UA and UPA are both used to describe, conceptually, the same type of occurrence or phenomenon. What may seem to be a lack of conceptual clarity can actually be viewed as a strength, where variation in socio-economic and environmental conditions allow for local innovation, offering 'best practice' examples of UA/UPA systems and policy approaches. Nonetheless, it is worth noting that the diversity of UA/ UPA production systems is a reflection of the various cultural, political, geographic, economic and climatic conditions where it is found. Such diversity in people, places and production systems has led to several useful definitions of UA/UPA. It should be no surprise that its complexity, purpose and benefits to practitioners and ecosystems vary in the global North and South. In this chapter, the broader term UPA will be used to discuss intra- and peri-urban food production systems.

Despite its 'age old origins', the UPA phenomenon did not emerge as a potential development tool until the 1970s. The World Commission on Environment and Development (WCED), or the 'Brundtland report', focused the world's attention on sustainable urban development and the potential role of 'urban agriculture', or UPA, in 1987:

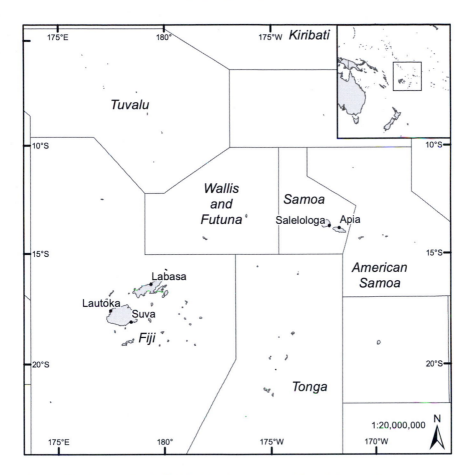

*Figure 13.1* Map of South Pacific Islands, including Fiji and Samoa. Base map: www.naturalearth.com

Officially sanctioned and promoted urban agriculture could become an important component of urban development and make more food available to the urban poor. The primary purpose of such promotion should be to improve nutritional and health standards of the poor, help their family budgets (50-70 per cent of which is usually spent on food), enable them to earn some additional income, and provide employment. Urban agriculture can also provide fresher and cheaper produce, more green space, the clearing of garbage dumps and the recycling of household waste.

WCED (1987, p254)

Despite its importance as an informal survival strategy and income source, attempts to formalize UA invite a host of constraints, both environmental (waste, pests, pollution) and institutional (land tenure, health, political). Overcoming these constraints requires consultation and collaboration amongst a range of stakeholders (community, civil society, government). Because urban agriculture is contributing to the nutrition and dietary diversity of urban poor households, some governments, recognizing its importance, are actively exploring ways to integrate UPA into formal urban planning policy (Dreschel and Dongus, 2010; Thornton, Nel and Hampwaye *et al.*, 2010).

As a holistic response to mitigate the impacts of global warming, urban agriculture is viewed as part of a low-energy ecosystems approach, where cities become more self-sufficient in food production, cutting consumption of fossil fuels and other inputs and reducing waste flows through reuse, recycling and composting (Marten, 2004; Roseland, 2005). Moreover, it is increasingly mentioned as a response to the impacts of increasing costs of fossil fuels on global food distribution, or 'food miles'. However, these debates largely take place in the global North, where 'locavores' and 'greener' consumer preferences for environmental and socially conscious products are part of a wider social movement seeking alternatives to globalized production of consumables, especially food. Through grass-roots movements and the development of community-supported agriculture concepts (Adam, 2006), the production, marketing and distribution of local food, including production from urban farms and community gardens, have played a role in formalizing urban agriculture in urban planning and sustainable agriculture policy in the western world (Hopkins, 2000). In developing countries, discussed in the next section, the status and stature of urban agriculture could not be more different.

## Urban agriculture in developing countries

In developing countries, the attitudes of former colonial governments towards UPA continue in recent times: UPA activities are often viewed as 'unsightly', they may be officially banned (Tinker, 1994, p5) and 'undervalued and resisted by generations of public officials' (Binns and Lynch, 1998, p778). However, some national and local governments are promoting UPA as a response to the challenges inherent to urbanization and economic liberalization, calling for self-reliance in food provision (Nel *et al.*, 2009). Urban population influxes have largely stemmed from rural small-scale farming households feeling the impacts of globalization and structural adjustment policies, which effectively ended protectionist, domestic economic policies in favour of open-market, export-oriented economy (see McMichael this volume). As a consequence, small-scale rural producers, struggling to adapt to these changes, often seek other means of employment and a better standard of living in cities. With limited prospects for formal wage employment, home ownership or secure land tenure, low-income and destitute urban dwellers often engage in urban agriculture for subsistence and income generation – or both (Thornton, 2009a).

Beyond its role as an informal response to unemployment and food insecurity, UPA is being discussed in culturally and geographically diverse places. For example, UPA is being considered as a formal strategy to stem the growth of peri-urban settlements through the inclusion of green space in Mexico City, Mexico (Torres-Lima and Rodriguez-Sanchez, 2008), and as a strategy for wastewater treatment in both Hyderabad, Pakistan (Van Rooijen *et al.*, 2010) and Accra, Ghana (Cofie and Adam-Bradford, 2006), where rapid urban population growth has led to innovative solutions in waste management. In these cases, UPA is proving, in practice, to be an adaptable system, tailored to suit local variables, needs and demands.

### Under what conditions has UPA thrived?

Since the 1950s, the dominant perception of 'development' has been a process that seeks to transform 'traditional' societies into 'modern' (Western) societies. This view has arguably left its indelible mark on generations of government officials and urban planners in developing countries, particularly in sub-Saharan Africa. The dualistic notion of cities as 'modern' centres of commerce and industry has effectively stigmatized urban food production as a distinctively anti-modern or 'backwards' activity. For most of the global South, therefore, UPA has been met with official disdain, indifference and outright banning of the practice in the latter decades of the twentieth century. Despite this, UPA has thrived in the global South, where post-war modernization strategies have failed to 'develop' the 'underdeveloped', as expressed by US President Harry Truman in his 1949 Inaugural Address.

Many of the oft-cited examples of thriving UPA are located in countries experiencing economic crises while undergoing structural adjustment. Cuba is a notable exception, where UPA has been part of a broader government strategy for local food production since the collapse of the socialist trading bloc in 1989 and the subsequent cessation of oil imports from the former USSR. This collapse led to a food and energy crisis, known as Cuba's 'Special Period in a time of peace' (Rosset and Benjamin, 1994). The lack of petroleum diminished the production and export earning capacity of Cuba's input-dependent monocultures (largely sugar cane), a situation that compromised the food security of the entire island nation and led to official support for UPA as part of a broader nationwide initiative of organic and sub-organic agriculture. Elsewhere in the south, many developing post-independence states borrowed excessively from foreign lenders in the 1970s, in part to finance public or social safety-net expenditures such as food subsidies. Structural adjustment programmes (SAPs), or austerity measures, created by the International Monetary Fund (IMF) and implemented by the World Bank (WB), sought to teach fiscal discipline to developing economies. SAPs require borrowing countries to liberalize their markets and to redirect social safety-net expenditures towards paying off foreign debt. A reduction of public expenditures often translated into the termination of consumer food subsidies. The demise of food subsidies during a transition to an open market has drastic impacts on the ability of the urban poor to meet their food and nutritional needs. As a local response, UPA expanded as an alternative strategy to ensure low-cost urban food supply amidst economic liberalization under SAPs. Its practice has been well documented as a response to food insecurity in countries negatively affected by neoliberal development strategies (Rogerson, 1993; Thornton, 2008).

Entering the 21st century, research on UPA's environmental and socio-economic impacts and supportive institutions has increased. This increase in research and support is testimony to not only the adaptability and resilience of UPA as a distinct production system, but also as a local response to cope with recurring crises brought about from global 'shocks' in the price and supply of food and fuel. Although it is valuable to reflect on the conditions where UPA has developed and thrived to understand how far it has come, it is more crucial to consider where it is going. The last decade has seen fewer governments holding onto the view of UPA as a rogue system, out of step with the post-war developmentalist view of the modern, urban-industrial city. Instead, and increasingly, governments are exploring ways to integrate UPA into official policy planning to mitigate the exploitative tendencies of an unpredictable capitalist system of food supply, access and availability.

### Factors contributing to vulnerability and risk

Discourses on food security have taken numerous turns that range from supply, price and consumption; entitlement and access; local production and distribution; livelihoods perspectives, and indigenous knowledge. A common thread among these shifts in thinking is the impact of the commoditization of food on populations in developing countries that exist on the fringes of the global free market. In 2006, the world witnessed sharp increases in the cost of food, when the food index grew by an average 57 per cent above 2006 levels (FAO, 2009). These increases corresponded with fears of chronic food insecurity and malnutrition in sub-Saharan Africa (SSA) – a region deemed highly unlikely to meet most of the MDG (Millennium Development Goal) targets by 2015. In other developing regions, concerns about persistent poverty in relation to food price inflation are rapidly emerging as a result of the impacts of recent natural disasters and the global financial crisis on food security and nutrition among populations in small island developing states (SIDS). In this chapter, these impacts will be explored in the South Pacific.

The impact of frequent and intensifying natural disasters as a result of climate change on food security has emerged as a primary concern. The implications of global warming on natural and human habitats on Pacific SIDS has been acknowledged at the United Nations (UN) World Food Summit and emphasized by UN-HABITAT (the UN Human Settlement Programme). The expansion and diversification of local food production has been recommended by the UN Food and Agriculture Organization (FAO, 2009), as a strategy towards sustainable livelihoods, where risk and vulnerability are managed through limiting dependency on imported goods by increasing local production. Coordinated efforts to integrate rural and urban production systems with the natural and urban ecosystems can effect secure livelihoods and adequate nutrition.

Arguably, South Pacific states are the most vulnerable of the SIDS 'fringe' economies, due to the combined effects of (a) remoteness from major markets, (b) susceptibility to increasing frequency and intensity of natural disasters related to climate change, and (c) natural resource limitations and susceptibility to crop diseases that restrict the potential to produce comparative advantages. Local level responses to global threats to food security in the south include local food production that incorporates urban food production or UPA.

## UPA in the Pacific Islands

In Pacific island countries (PICs), urbanization and expanding squatter settlements are important development issues with attributes that are unique to the wider context of the 'global South'. These unique traits include the fact that the development of towns and cities in the South Pacific islands is quite recent, only emerging from the post-colonial era of the 1960s. Small island economies are also characteristically isolated from large export markets; this constraint is aggravated by the vulnerability of key sectors, such as agriculture, to the effects of climate change. However, development discourse tends to incorporate the capacity of SIDS to adapt to Western economic models. This view has been criticized as a deterministic understanding of the Pacific SIDS as 'islands in the ocean', existing on the fringes of the globalized world, at risk of disruptions to the flow of goods and services, and exposed to price fluctuations (external 'shocks') as determined by global markets. This view is in contrast to an indigenous interpretation of 'an ocean of islands' (Barnett and Campbell, 2010, p49), in which Pacific communities are seen as highly-skilled resource managers, rich in local knowledge, and with the capacity to find and implement localized solutions to mitigate the worst that global change may bring.

The disruptive impacts of global change on local food supply networks have received limited attention in the largely coastal urban areas in Pacific SIDS that are prone to salination of field crops, fruit orchards and home gardens by frequent and intense natural disasters. These networks upon which villagers are becoming increasingly dependent for food and income include subsistence and market food (and forestry) production from urban environs and urbanizing areas (including villages near expanding towns). Like other developing countries experiencing rapid urbanization, national and local governments in PICs have limited resources to cope with burgeoning urban populations and their subsequent demands for service provision, housing and employment (Connell and Lea, 2002). There is also limited understanding, at governmental levels, of the importance of the urban ecosystem to sustainable urban-based livelihoods in rapidly urbanizing PICs.

In comparison to sub-Saharan Africa, the literature on urban agriculture as a survival and coping strategy in the Pacific islands has not received as much attention as it deserves. Reasons for this could include that, among many PICs, human development indicators (available from the United Nations Development Programme, Human Development Report) – life expectancy at birth, literacy, maternal health – are near levels that compare favourably to developed countries. Nonetheless, over the years a number of important studies have been carried out in the region looking at UA and related topics concerning urban subsistence food gardening and urban agroforestry (Thaman *et al.*, 2006; Thornton, 2009b), small-scale farmers, and local food production (Ward, 1959; Crocombe 1987). In contrast to the neoliberal development approach – that emphasizes monoculture production systems for export agriculture – UA, as part of a local food production system, is suggested as a more effective strategy for small-scale farmers. It is perhaps better suited to the environmental conditions common to Pacific islands (fragile, yet arable soils) and can offer local market stability to producers who have turned away from inequalities inherent in commercial export agriculture due to difficulties in achieving comparative advantage, and who have been victimized by fluctuations in the global market and costly inputs.

Industrialization, rural-urban flows, resource exchanges and transition of rural agricultural land into peri-urban zones can vary among South Pacific island states. In the following sections, issues of urbanization, poverty and local food production are discussed in Samoa and Fiji. These Pacific island countries share some similarities, in terms of maintaining their traditions and culture, their strong Christian beliefs and the allure of their tropical environments. Although the levels of urbanization and responses to food insecurity are different, they both offer useful lessons to the region in coping with similar problems through alternative, bottom-up approaches. At the local level, where grass-roots responses emerge, small-scale and community-based initiatives are influencing official policy making. This is not surprising as, worldwide, towns and cities can (and do) pursue and implement alternative solutions to urban problems that may not reflect broader national policy.

## Case study: local food production in Samoa and Fiji

### *Samoa*

Samoa was the first PIC to regain its independence in 1962 and is currently experiencing rapid urban growth in Apia, the nation's capital, on Upolu Island. Samoa's population (180,000) is unequally divided between the two largest islands of Upolu (76 per cent) and Savai'i (24 per cent). On Upolu, the lines differentiating 'urban' from 'rural' food production are not entirely visible due to the fluid nature of movements and flows of people, food and other resources between rural and urban households and markets. As urban expansion continues, UPA as a distinct system will become clearer. Moreover, perhaps former colonial industrial areas near Apia represent newly contestable peri-urban zones. In any case, this urbanization trend is visible on Samoa's more traditional or 'rural' island, Savai'i, where the harbour village, Salelologa, is the focal point for inter-island transfers of goods and services, and provides a busy marketplace for intra-island trade. It is also the first port-of-call for tourists visiting the island. Salelologa is being developed as a secondary growth node (or town), to service the needs of the island's residents and to stem the inter-island flow of rural-urban job seekers to Apia. The Apia urban area includes a collection of surrounding villages, comprising 52 per cent of Upolu's population. Although each village has its own *matai* (chief led system) of governance, social structure and customary land tenure, they are evolving as part of an emergent Apia 'metro' district. Like other South Pacific island states, subsistence agriculture is the overarching household activity in the villages. Much of the local food and non-food production is sold at Fugalei Market, in Apia.

Although poverty reduction policies do not currently exist in Samoa, the government has recognized increasing economic hardships for many households. In the urban areas, food security problems have been identified as a growing issue. Over one-half of all reported cases of childhood malnutrition are from within greater Apia (Muagututi'a, 2006, p62). The most vulnerable groups in Samoa include the urban poor with limited land access, the rural poor who lack cash income, and the young. One of the key challenges for the Samoan economy is its dependence on remittance income, derived from the significant numbers of its population living overseas. Estimates place such remittance income in the order of 20 per cent of GDP. By comparison, agriculture represents 6.7 per cent of GDP. While agriculture is only a minor contributor, in a rural subsistence economy, its role as a survival strategy – as opposed to a creator of wealth – is critical.

In contrast to many other Pacific island states, the *faa Samoa* (Samoan way of life) has endured, despite external pressures to modernize social, political and economic life (Thornton, Kerslake and Binns, 2010). At one level, this has ensured the continuance of communal-based land ownership, but, at another level, this is at odds with global expectations of land reform and privatization. Recently, since the mid-1990s, various regional trade agreements in the Pacific, IMF structural adjustment policies, commitment to the Millennium Development Goals (MDGs) and (anticipated) entry into the World Trade Organization (WTO) are forcing change. Rather than increasing liberalization, as one might anticipate with WTO membership negotiations, emphasis is placed on the development of niche market food production (by the Central Bank of Samoa) and on local food production for urban markets.

### Potential for developing local food systems

Recognizing the risks and vulnerability of deepening trade liberalization,[2] the Government of Samoa is promoting small-scale farming for domestic food security, as well as niche agriculture with export potential. However, the success of any agricultural renaissance might hinge on the willingness of Samoans to continue engaging in farming, and to clear land for new commercial plantations. In an earlier study of Samoa's banana industry, Ward (1959, p126) observed that the existence of uncultivated land was attributed to the 'low social status accorded to skill and diligence in agriculture'. This view is mostly related to the traditional division of labour in the village, where a *matai* (head of family, or chief) will assign *taulele'a* (untitled men) to work in the plantations (a commercial holding or garden). The historical tenure system perpetuates traditional values while also ensuring subsistence livelihoods, but also has the negative associations of the lack of access to credit or potential foreign investment. In addition, the future of the agricultural economy has been negatively impacted by natural disasters and loss of labour to seasonal work opportunities in New Zealand and other countries.

During interviews, agricultural officials were very cautious in contemplating a future in which Samoa might have niche-market farmers or producers of high-quality products for the local market. Despite this, the current Samoan Development Strategy seeks to support small businesses, cottage industries, vegetable production, and taro and banana chips initiatives. The Ministry of Agriculture views growth in local agriculture as a means to ensure food security and increase commercial activity. There is also a clear desire on the part of government to increase agricultural employment and to promote private sector activity in the areas of copra, coconut oil and research into developing biofuels. The ambitions of a nascent biofuels market, however, are more regional, than global. Although still in its experimentation stage, the main objective of this government-supported initiative is to lessen vulnerability to petrol/diesel price-shocks through innovation, using locally-resourced feedstock and the knowledge economy, to meet a degree of the energy demands of a growing population and burgeoning tourist sector.

*Development from below: niche market production*

In the early 1990s, a group of Samoan women formed Women in Business Development Incorporated (WIBDI), in response to devastating cyclones, economic change, limited income opportunities for rural women and the desire of the Samoan Ministry of Women's Affairs to support this vulnerable sector. Three concepts are critical in WIBDI's activities: namely, weaving together technology, trade and tradition, and in so doing helping the more disadvantaged to use their talents, generate cash income and respect and use tradition. Current projects include fine mat weaving, honey production, organic farming and virgin coconut oil production.

Movement into organic farming has been difficult to achieve, as it requires areas of cultivation to be certified as having remained out of production for several years before being declared organically certified. The overlap with national interest in this approach has been secured through forming links with the Samoa Organic Farmers Association. Certification by the National Association for Sustainable Agriculture, Australia (NASAA) for organic extra virgin coconut oil, which commands a premium price, has opened markets in Australia, Germany,

New Zealand and the USA. To facilitate coconut production and export, the Pure Coconut Oil Company was established in 1990 to encourage the setting up of micro-production plants using 'direct micro-expelling technology' to provide cold-pressed oil from dry coconut meal. In 1996, five villages on the two main islands were operating as coconut oil production centres. The majority of this production is sold directly to the Body Shop (UK) for hygiene and cosmetic products. In addition, efforts have been made to secure Fair Trade registration. Other organically certified crops that are exported include bananas, coconuts, cocoa, nano juice and ginger. To date, 200 farms have been organically certified (out of 14,800 economically active households), 100 are being processed and 200 are on the waiting list, indicating the growing popularity of this approach. In terms of the future, the organization is working with parallel groups in Tonga, Fiji and Niue to pool resources and reduce risk through having a regional approach.

Like other Pacific SIDS, agriculture in Samoa has been a crucial source of revenue (banana, copra) and frustration (cyclones, infestations, distance). As economic trade agreements – vestiges of old colonial trade routes – fade away, Pacific SIDS must adapt and recognize constraints and opportunities brought by economic globalization and climate change. In Samoa, opportunities exist in innovations in niche market products (virgin coconut oil) and biofuels experimentation. The constraints include mitigating the impacts of global warming on agri-based industries, where a regional approach to ensure supply of key resources seeks to minimize risk. However, developing new industries takes time, while population growth, urbanization and limited job prospects place immediate demands – none more pressing than a varied, nutritious and affordable food supply. Government, non-governmental and community-based organizations continue to promote the importance of local food production for sustenance and urban markets. It may be that Samoa's geographic isolation, in addition to its social and cultural integrity (including land tenure systems), which is widely acknowledged as unique among the Pacific island states, has effectively maintained the enduring role that rural and urban food production plays in providing low-cost food, including culturally important staple crops. In the following section, the development of UPA in Fiji, an economic driver among the Pacific SIDS, is following a path similar to many African countries. Poverty and squatter settlement expansion in Fiji is highest amongst South Pacific SIDS. At the level of local government, UPA is recognized as a legitimate component of a wider strategy for poverty alleviation, nutrition and food security.

### Fiji

Fiji is comprised of 322 islands, 110 of which are inhabited. The main islands of Viti Levu and Vanua Levu account for 85 per cent of the total land area. Currently, Fiji has an overall urban population of around 50 per cent, which is higher than the estimate of 40 per cent for the Pacific island region. Fiji is categorized as a middle income country, with a per capita income of 2,172 USD.

The contribution of agriculture to GDP is estimated to be around 10.49 per cent (2008), with sugar and subsistence farming dominating the sector. In 2006, the Human Development Index (HDI) ranked Fiji relatively high in various human development indicators (life expectancy, adult literacy, quality of life). Compared to other developing countries, Fiji received an HDI rank of 90 out of 178 countries, which represents the highest ranking amongst the small PICs (Samoa, HDI: 75; Tonga, HDI: 55).

However, in parallel with rapid urbanization, squatter settlement growth has 'mushroomed' in Fiji. The number of 'registered' squatter settlements in Fiji range from 184 to 190, which have a combined squatter population of 90,000 to 100,000, respectively. Over a seven year period (1996–2003), the squatter population in Fiji has risen 73 per cent, as the 'first wave' of land leases, signed 30 years ago, have begun expiring (Thornton, 2009b). Nationwide, the squatter population consists of 46 per cent landless indigenous Fijians and 53 per cent ethnic Indian (Indo)-Fijians.

The impact of the garment industry's decline and its impact on women in particular are often overlooked in discussions of poverty and squatter growth in Fiji. Thousands of women, mostly young Indo-Fijian garment workers, lost their jobs in the economic downturn that followed the 2000 putsch. In seeking alternatives to unemployment, many of these women are now believed to be engaged in UPA-related livelihoods. In addition to ethnic and gender biases, class inequalities also exist, as not all indigenous Fijians reap the benefits of land ownership. Land ownership is reserved for high-ranking members of the indigenous *mataqali*. Hence, the issue of landless Fijians is not confined to the ethnic Indo-Fijian minority, or evicted sugar cane farmers.

*Lautoka City Council: making inroads to a formal*
*peri-urban agriculture sector?*

Many sugar cane farmers on the west coast of Viti Levu live in established residential farming communities in Lautoka, Fiji's second largest city (population 43, 000). In one such community, 'Field 40', Indo-Fijian households with expired agri-leases are considered by the Lautoka City Council (LCC) to be 'technically squatting', as they do not have a land title or no longer have consent from a land owner – through the Native Land Trust Board (NLTB) – to reside on indigenous land. In the 1990s, due to concerns over increasing poverty and crime, the LCC applied for boundary extensions to upgrade service provision in settlement areas that had extended beyond the city boundary. Following a 15-year process, the boundary extensions were finally approved in 2006, though by this time the number of squatter settlements, resulting largely from expired agri-leases, had expanded significantly and could no longer be contained even in the extensions applied for 15 years earlier (Thornton, 2009b). More successful is the LCC's urban food market initiative: a significant statement that embeds urban food production as an important contributor to the city's urban food supply chain. In this regard, urban food production in Lautoka reflects trends in other cities and towns in the global South and North, where UPA is an important source of local food and household income for many urban residents.

Countrywide, indigenous Fijians produce 15–17 per cent of food in urban areas, while Indo-Fijians produce around 10 per cent of urban food. The availability of low-cost food at urban markets is sourced, in part, from low-income producers: a relationship which provides needed sources of income and food security for the poorest and most vulnerable households. The LCC has recognized this contribution and facilitated the growth of UPA, as evident in the establishment of urban market stalls. The urban market in downtown Lautoka offers a sheltered area and refrigeration services for vendors of meat, fish, vegetables and fruits to operate. Although it is comparable in size to the Suva market (nearly 60,000 sq. m.), the Lautoka urban market operates in a covered formal structure. A market licence is required to help cover the costs of maintenance, as well as to ensure quality control for the LCC's health and safety requirements – the cost of a market licence fee is F\$2.20 per day or F\$100 for one year. Finally, the Public Health Act in the City of Lautoka is supportive of urban agriculture as an acceptable source of household food. The Fijian government, like the Samoan government, seeks to enhance local food production through the promotion of self-sufficiency, including UPA, as a strategy to mitigate risk from price shocks in staple foods, oil and the impact of global warming on the availability and access to locally favoured and culturally important foods.

Increasing economic hardship is a fact of life in Pacific island countries; a notion that might run counter to a Western perception of the South Pacific islands as places of tropical tranquillity and socio-economic equality. The reality is that human and natural ecosystems in the Pacific islands are experiencing the negative impacts of economic globalization and climate change, which is threatening fragile habitats, human settlements and centuries of regional diversity in indigenous knowledge, traditions, customs and livelihoods.

## Conclusion

This chapter discussed initiatives for local food production, which includes grassroots urban production systems (Fiji) and civil society groups in partnership with small-scale farmers (Samoa). In countries categorized as 'small', distinctions between what constitutes urban, in contrast to rural, production are perhaps not as important as the integration of these systems as elements of a local food production network. The importance of local food production is increasingly recognized at government levels as a strategy to strengthen food security and local economies. Public policy directly supporting UPA systems is a further reflection of the significance of its contribution to household food security, savings and source of income. For the global South in particular, consumption of locally-produced food through UPA can fluctuate from environmental, economic and socio-political conditions. During the ten-year civil war in Sierra Leone, for example, UPA was a food-source lifeline for urban dwellers. In post-apartheid South Africa, in contrast, social grants are contributing a consistent source of income that effectively suppresses the emergence of a robust UPA sector, as compared to other African cities such as Lusaka (Zambia) and Freetown (Sierra Leone).

In the South Pacific islands, subsistence agriculture is a major source of food, providing up to 80 per cent of food for household consumption. In-depth studies on consumption of food from UPA sources are very few and those that exist are past their prime. However, discussions with consumers at urban markets (from the case studies) provide some indication that consumption of locally produced food, including that sourced from UPA, can range from 15–60 per cent; the higher figure a result of increasing prices in basic foods. Here also, there is real potential for nascent industries to exploit niche markets in agriculture, as demonstrated by virgin coconut oil and nonu juice export production. Although some level of local ownership exists, the extent of participation in decision-making from village level producers is not clear.

The emergence of a dialogue within Pacific SIDS to shift focus away from exogenous sources to endogenous solutions could undoubtedly lead to criticisms of protectionist policies. However, the seemingly unrelenting onslaught of crises borne out of Western economic globalization and climate change in 'fringe economies' and 'isolated states' is likely to increase the appetite of SIDS to develop local food production and regional solutions for resource management as a means to mitigate threats to food security and social and economic development. A sense of urgency seems to be emerging at the level of civil society and local government that a domestic food supply network integrating urban and rural producers must no longer be undervalued at the national level. The capacity of developing countries to ensure their own equitable and accessible supply of food, to nurture human capital, will determine their potential for social and economic development.

## Notes

1  Global change, or economic globalization and climate change, is a broad reference to processes occurring at a global scale that have consequences on all scales
2  The Government of Samoa is negotiating with the World Trade Organization (WTO), its terms of membership. Part of this process hinges on Samoa graduating from its current status of 'less developed country' (which brings beneficial trade concession) to 'developing country' status (the loss of trade concessions).

## References

Adam, K. (2006) 'Community supported agriculture', ATTRA, National Sustainable Agriculture Information Service, http://attra.ncat.org/attra-pub/PDF/csa.pdf, accessed 15 June 2010

Barnett, J. and Campbell, J. (2010) *Climate Change and Small Island States*, Earthscan, London

Binns, T. and Lynch, K. (1998) 'Feeding Africa's growing cities into the 21st century: the potential of urban agriculture', *Journal of International Development*, vol 10, pp777–93

Cofie, O. and Adam-Bradford, A. (2006) 'Organic waste reuse for urban agriculture', in R. van Veenhuizen (ed) *Cities farming for the Future: Urban Agriculture for Green and Productive Cities*, International Development Research Centre, Ottawa

Connell, J. and Lea, J. (2002) *Urbanization in the Island Pacific*, Routledge, London

Crocombe, R. G. (ed) (1987) *Land tenure in the Pacific*, 3rd edn, Institute of Pacific Studies, Suva, Fiji

Dreschel, P. and Dongus, S. (2010) 'Dynamics and sustainability of urban agriculture: examples from sub-Saharan Africa', *Sustainability Science*, vol 5, no 1, pp69–78

FAO (Food and Agriculture Organization) (2009) 'The state of agricultural commodity markets: high prices and the food crisis – experiences and lessons learned', FAO, United Nations, Rome

Hopkins, R. (2000) 'The food producing neighbourhood', in H. Barton, (ed) *Sustainable communities: The potential for eco-neighbourhoods*, Earthscan, London

Marten, G. (2004) *Human ecology: basic concepts for sustainable development*, Earthscan, London

Muagututi'a, S. R. (2006) 'The Human Development Indices', in A. So'o, U. Va'a and J. Boon (eds) *Samoa National Human Development Report*, chapter 4 p47–68, National University of Samoa, Apia

Nel, E., Hampwaye, G., Thornton, A., Rogerson, C. and Marais, L. (2009) 'Institutional responses to decentralization, urban poverty, food shortages and urban agriculture', *GDN Working Paper Series*, Working Paper no 36, pp1–29, Global Development Network (GDN)

O'Brien, K. and Leichenko, R. (2003) 'Winners and losers in the context of global change', *Annals of the Association of American Geographers*, vol 93, no 1, pp89–103

Ponter, B. (1975) 'Rural migration in the South Pacific', in J. Hardaker (ed) *The subsistence sector in the South Pacific*, University of the South Pacific, Suva, Fiji

Rogerson, C. M. (1993) 'Urban agriculture in South Africa: policy issues from the international experience', *Development Southern Africa*, vol 10, pp33–44

Roseland, M. (2005) *Toward sustainable communities: resources for citizens and their governments*, New Society Publishers, Gabriola Island, BC

Rosset, P and Benjamin, M. (eds) (1994) *The Greening of the revolution: Cuba's experiment with organic agriculture*, Ocean Press, Melbourne

Thaman, R., Elevitch, C. R. and Kennedy, J. (2006) 'Urban and homegarden agroforestry in the Pacific islands: Current status and future prospects, in B. M. Kumar and P. K. R. Nair (eds) *Tropical Homegardens: A time-tested example of sustainable agroforestry*, Springer, Dordrecht, Netherlands

Thornton, A. (2008) 'Beyond the metropolis: Small town case studies of urban and peri-urban agriculture in South Africa', *Urban Forum*, vol 19, no 3, pp243–262

Thornton, A. (2009a) 'Pastures of plenty?: Land rights and community-based agriculture in Peddie, a former homeland town in South Africa', *Applied Geography*, vol 29, no 1, pp12–20

Thornton, A. (2009b) 'Garden of Eden?: The impact of resettlement on squatters 'agri-hoods' in Fiji', *Development in Practice,* vol 19, no 7, pp884–894

Thornton, A., Kerslake, M. and Binns, T. (2010) 'Alienation and obligation: religion and social change in Samoa', *Asia Pacific Viewpoint*, vol 51, no 1, pp1–16

Thornton, A., Nel, E. and Hampwaye, G. (2010) 'Cultivating Kaunda's plan for self-sufficiency: is urban agriculture finally beginning to receive support in Zambia?', *Development Southern Africa*, vol 27, no 4, pp613–625

Tinker, I. (1994) 'Foreword: Urban agriculture is already feeding cities', in G. A. Egziabher, D. Lee-Smith, D. G. Maxwell, P. A. Memon, L. J. A. Mougeot, and C. J. Sawio (eds), *Cities Feeding People: An examination of urban agriculture in East Africa*, International Development Research Centre, Ottawa

Torres-Lima, P. and Rodriguez-Sanchez, L. (2008) 'Farming dynamics and social capital: a case study in the urban fringe of Mexico City', *Environment, Development and Sustainability,* vol 10, no 2, pp193–208

Van Rooijen, D. J., Biggs, T. W., Smout, I. and Drechsel, P. (2010) 'Urban growth, waste-water production and use in irrigated agriculture: a comparative study of Accra, Addis Ababa and Hyderabad', *Irrigation and Drainage Systems*, vol 24, nos 1–2, pp53–64

Ward, G. (1959) 'The banana industry in Western Samoa', *Economic Geography*, vol 3, no 2, pp123–137

WCED (World Commission on Environment and Development) (1987) *Our Common Future*, Oxford University Press, Oxford

# 14

# CONCLUSIONS: TOWARDS A MORE JUST AND FLEXIBLE GLOBAL FOOD SYSTEM

*Christopher Rosin, Paul Stock and Hugh Campbell*

### Thinking the unthinkable: what kind of crisis is this?

It seems appropriate at this point to once again refer to the title of Tim Lang's (2010) article: "Crisis? What Crisis?" Despite its beginnings in a conference designed to examine the global food crisis from a New Zealand perspective, the food crisis as such does not play a prominent role in most of this volume's chapters. In attempting to identify the root causes of the 'Crisis of 2008', the general consensus of the contributors to the 44th Otago Foreign Policy School (OFPS) was that we were facing an ongoing problem caused by the persistent social, political and economic structures which limit the capacity of our global society to effectively reduce the incidence of hunger. Many of our contributors furnished evidence for this in the everyday experiences of food production in particular places that are subject to diverse local pressures as much as global food commodity prices. All of which raises the issue of whether the food crisis was a temporary blip or a sign of things to come.

The discussions stimulated by the participants in the OFPS clearly established the coalescence of new dimensions to the problem of food insecurity. The emergence of biofuels as a competitor to food crops is an unwelcome new pressure for securing food provisioning in many countries. The new dynamics of financialization have created markets that trade in abstracted food futures (and encourage broader patterns of unhelpful investment in food-related economic activities). The consequences of the further development of this process warrant close examination. However, the overall conclusion of discussions at the OFPS (and the chapters in this volume that have followed) is that much of what we are seeing (in 2008 and now again in 2011) is not a novel event. Moreover, to call it a crisis actually obscures the long term trajectory that has led to this current level of failure in parts of the world food system.

Putting this another way, the most profound insight of the OFPS was recognition of the extent to which once marginal or radical ideas related to feeding the world have crept into the mainstream. This encroachment further raised doubts about the wisdom of our overall global policy direction towards hunger and the role of market-driven countries such as New Zealand. This development was fostered by the conference format which provided a forum for addressing some uncomfortable political realities that remain unthinkable in a highly neoliberalized country like New Zealand. The conversations that emerged centred on the need for serious transformations of the neoliberal, market-driven model for providing solutions to world hunger (which we have referred to throughout this book as the 'business-as-usual' model). As a training opportunity for the new intake of New Zealand's government officials, diplomats, and policy specialists, the OFPS did not represent the typical meeting place of the coalition of interests (including academics, NGOs, a broad coalition of groups across the Developing World, and a range of other critics of the neoliberal project) that have long questioned the potential of market-solutions to solve key challenges such as world hunger. It was the discourse emerging from the former group that would probably leave the remaining apostles of neoliberalizm with the greatest cause for concern. The OFPS enabled a broad coalition of potentially influential political actors in New Zealand to think the unthinkable: was the neoliberal model broken and, if so, what replaces it? It was the importance of that question that led the editors of this volume to draw together the insights (and responses) of the significant group of experts who had attended the 44th OFPS.

In this concluding chapter, we will briefly summarize the case made both at the OFPS and in this book for why the current food crisis was not a blip, but an important signifier that the business-as-usual model is unravelling. We will then address the range of answers that have been proposed for reorienting the global food system in light of the apparent failure of the neoliberal model. We acknowledge the potential of these recommendations to contribute to the global capacity of agriculture to meet increasing food demand, but also challenge the transformative potential of practice, technologies and policies that fail to address underlying systemic problems of food security. In the final part of the chapter we return to the concept of utopias raised by Stock and Carolan in the first section of the book, arguing that any shift to a more just and flexible food system will require a new framework for engaging with global food. What does an ideal global food system look like? How do we imagine the balancing of extremely diverse expectations? Such a framework cannot rely solely on market logics nor fall into the category of radical or inflammatory if it is to inform more appropriate policy and governance interventions.

## Understanding the failure of 'business-as-usual' in the world food system.

Through the process of identifying pernicious (and persistent) elements of global food relations, the authors in the first section of this collection draw our attention to underlying patterns that prevent us from feeding the world. In a review of the relations of production (both social and environmental) in the global food system, Jules Pretty emphasizes the complexity of factors contributing to the current situation of global food insecurity. He further attributes the limited efficacy of our collective response to a failure to acknowledge and respect the socio-ecological interactions underlying food production. Hugh Campbell, on the other hand, notes the particularly counterproductive subjugation of food policy to the concerns of domestic and international political aspirations. In the domestic context, agriculture policy has been about securing incomes for domestic farm sectors. Abroad, under the guise of 'aid', food is first treated as a quantity – the surplus production of an overheated farm sector – and then as a weapon – cajoling cooperation from friendly governments, punishing the enemy by withholding it. Bill Pritchard details a similar issue in relation to the unwavering bias to neoliberal strictures on the governance of international trade. In these situations food becomes a mere commodity, a thing to be bought and sold with no recognition of its essential contribution to the survival and health of humanity. According to Philip McMichael, the implications of such a treatment of food are apparent in the impact of biofuel development in the 2008 crisis – as just a commodity. The value of maize was determined, not by its role as a staple and nutrient-rich food, but by our thirst for fuel to drive vehicles. As if this evisceration of food were not sufficiently abhorrent, McMichael demonstrates the further obscuring of its essential dietary nature through the trading of food commodity futures, valued solely as capital investments to be bought and sold for profit.

The remaining contributions to the first section of the book demonstrate how this shift in the nature of food has had significant impacts on global engagement with human rights, health and aspirations for a better society. For example, Claire Mahon documents that, despite international commitments to human rights, including the right to food, too many of the world's population remain hungry. Colin Butler and Jane Dixon examine the parallel emergence of a reductionist approach to nutrition in which food becomes a quantifiable input to human health. They argue that the human preference for high energy foods and to gorge on excess, as well as the metabolic rift that obscures the impacts of food production on environment and society, contributed to the subsequent replacement of diverse diets with those dominated by a limited set of high-energy foods and synthesized nutritional supplements. Stock and Carolan illustrate the implications of the underlying quantification of the global food system on shared expectations of the achievement of a better society. The utopian aspiration to supply a sufficient quantity to feed the world detracts from the qualities of food, which they argue can contribute to utopian perspectives that better acknowledge the moral imperatives of accessible and healthy food for the world. We will return to this point later in this conclusion.

In summary, the first part of the book outlines the case for why the business-as-usual model is not working. The combination of political manipulation of food and the cultural transformation of everything into numbers hides our collective inability to feed the hungry. The final, and possibly most damaging, aspect is the belief that food should be cheap – squeezing farmers at home and abroad but also underwriting the profitable expansion of a disenchanting, commoditized global market for food. One of the longest, consistent trajectories since the Industrial Revolution has been the decreasing proportion of household income spent on food – economists call this Engel's Law. Commencing with around 85 per cent of poor-household income being spent on food in the 1880s, the slow and continual cheapening of the relative cost of food to households in the Developed World plateaued only recently, with the US reaching a low point of around 10 per cent of income being spent on food. Yet, the underlying truth of Engel's Law is that food just can't keep getting cheaper forever. If we have reached the point of reversal of Engel's Law, this will prove to be devastating for many aspects of the business-as-usual model in the world food system.

While the first part identifies a poorly functioning food system, the second begins to identify the opportunities for change and the constraints to alternative food production practices. If the dominant business-as-usual model is broken, is there one alternative model or a range of options waiting to replace it? None of the authors in this part would argue that their case study demonstrates a singular solution to what ails the food system at a global level. Each does, however, offer insight to the challenges faced particularly at the site of production in the global food system.

In their chapter, Geoff Lawrence, Carol Richards, Ian Gray and Naomi Hansar paint a pessimistic picture of local response and adaptation to the changing climate in the Murray-Darling Basin of Australia. Despite the increasing severity and duration of drought in the region, a deep-seated focus on productivism in Australian agriculture and culture more generally appears to overwhelm calls for more sustainable practice. The first challenge, in their case, is finding credible pathways for enacting alternatives to productivist agriculture. Jeff Nielson and Bustanul Arifin use the example of rice production in Indonesia to illustrate the potential for political expediency and privilege to displace concerns regarding access to food through a myopic focus on production. Their critique is muted to some extent, however, by the apparent benefits of self-sufficiency policies in the context of a food crisis driven by escalating commodity prices – a situation indicative of the complexities of domestic and global food systems. The case studies from Africa and South America interrogate the impacts on livelihoods and social relations as food producers in developing countries seek to benefit from access to wealthier export markets. Navé Wald, Christopher Rosin and Doug Hill examine a more threatening situation in which an export crop (soybeans) favours large-scale production to the detriment of existing *campesino* production and the environment. In the case of Argentina, the importance of a high-value agricultural export to the legitimacy of the national government interferes with the actions of the *campesinos*, who have adopted the rallying cries of food sovereignty in their efforts to defend the tenuous rights to the land. Kiah Smith and Kristen Lyons find that production for niche export markets provides both benefits and social disruption for small-scale production in Kenya and Uganda. Whereas such markets can act as a means to encourage a greater focus on food qualities, they also expose producers to social and economic valuations that may be ill-suited to local cultures. In the final chapter, Alec Thornton provides a more optimistic case study of the adaptation of small-scale producers to increasing urbanization in Samoa and Fiji. The ability to engage in urban agriculture, selling to both the domestic and tourism markets, provides an economically viable production alternative.

Together, the case study chapters demonstrate the capacity that the ideal of global food security has to impact local food production systems and the daily lives and survival strategies of rural populations. The ideology of productivism has long served the global community through the rapid escalation of food quantities in the face of equally rapid increases in demand. As the cases of Australia and Indonesia demonstrate, however, this ideology is increasingly challenged on the basis of the environmental and social externalities of production-driven food systems. The ideology of the free market represents a further ideal that often reinforces productivism. The remaining case studies illustrate the influence of the market ideology, albeit in distinct contexts. The focus on export production in Africa and Latin America negatively impacts both the cultural values and social rights of domestic producers. In the South Pacific, the market provides a less contested outcome, albeit the producers are largely engaged only in domestic trade.

As a whole, the case studies provide compelling evidence as to why a single solution to world hunger has been so hard to achieve. Each of the cases describes the potential adaptive capacity of local producers, yet the underlying concern of the contributing authors is that more just, flexible and productive food systems are subject to the overwhelming influence of structural constraints and local context. Perhaps the key conclusion to be drawn from this group of cases is that we must abandon the beguiling notion that there is *one* solution for world hunger. This, after all, is a significant part of the ideological power of the neoliberal market model – the claim that one size does fit all. In addition, this model suggests that, at a global level, politicians, policymakers, institutions and societies can be progressively disciplined for solving the ills of the world. The insights from this volume open up three key areas requiring further discussion in the remainder of this chapter. First, what happens once we realize that there is no 'silver bullet'? What governance options do we put in its place? Second, how do we address the fundamental ideological and cultural problem in how we understand food? Finally, can we bridge the gap between production and profit on one hand, and healthy consumption and abject hunger on the other? How do we even begin a dialogue on what that bridging might look like?

## Formulating responses 1: options and governance

A wide range of well-informed recommendations for achieving food security already exist – including those from joint, cross-disciplinary projects such as the IAASTD (International assessment of agricultural knowledge, science and technology for development) (McIntyre, *et al.*, 2009) and the UK Food Security Review (Godfray, *et al.*, 2010) to more narrowly focused scholarly assessments, to journalistic positions, to organizational planning. Many of these inform the solutions included in Jules Pretty's chapter. In Table 14.1, we further catalogue a selected set of recent recommendations for reconfiguring the global food system. We have categorized these on the basis of their capacity to initiate or contribute to changing business-as-usual in the global food system. In making this distinction, we are not arguing that the recommendations listed in the right-hand column are to be avoided or abandoned – the objectives they are intended to achieve likely

Table 14.1  Solutions to the food crisis as identified in selected sources

| Systemic Change | Non-Systemic Change |
| --- | --- |
| Alternative food production systems: <br> • 'green' agriculture[l] <br> • agroecology[e] <br> • sustainable intensification[f,i,j] <br> • farms as ecosystems, with focus on the internalization of externalities[h] <br> • waste reduction[b] <br> • multifunctionality in agriculture[h,m] <br> • recognition of women in agriculture[h,i] | Food production systems: <br> • agronomic science[d,f,i,j] <br> • genetic modification of crops/animals[j] |
| Alternative food consumption systems: <br> • certification programmes that make quality and origin known[a] <br> • social movements promoting food quality[a] <br> • sustainable diets[g,k] | Food consumption systems: <br> • infrastructure investment[l,m] |
| Alternative food governance systems: <br> • World Food Bank[b] <br> • integrated approach to food security, sustainability and agricultural knowledge, science and technology[f,h] <br> • focus on poverty and livelihoods, environment, health and nutrition, equity[h] <br> • sustainable food social movements[a] <br> • 'ecological public health'[g] | Food governance systems: <br> • reform existing structures like CGIAR and WTO[c] <br> • focus on food security and investments[h] <br> • free trade[l,m] <br> • population control[b] |
| Alternative food valuation systems: <br> • food as values[g] <br> • imagining sustainable food systems[a] | Food valuation systems: <br> • food futures trading[m] |

Notes:

a  Blay-Palmer (2010)
b  Brown (2011)
c  Clapp and Cohen (2009)
d  Cribb (2010)
e  de Schutter (2010)
f  Foresight (2011) and Godfray, *et al.*, (2010)
g  Lang, *et al.*, (2009)

h  McIntyre, *et al.*, (2009)
i  National Research Council (2010)
j  Royal Society (2009)
k  Sustainable Development Commission (2009)
l  UNEP (2011)
m  World Bank (2008)

remain necessary aspects of meeting the increasing demand for food. The point is that, on their own, these recommendations can only ever perpetuate the business-as-usual model that has proven its vulnerability to a range of perturbations, as evident in the food crisis of 2006–2008.

By contrast, what we have classified as more transformational recommendations all involve significant shifts in the structure and operation of the global food system, such that we necessarily begin to treat food not as a mere quantity. The recommendations for alternative production systems involve both environmental and social aspects of food production. Many of these are offered by long-time proponents of more sustainable and ecologically sound production systems. Such changes involve the recognition of environmental limits and working with, rather than imposing technological controls over, the ecosystems in which food is produced. In addition, there is the need to acknowledge the multiple outputs of agriculture and the multiple roles – especially those of women and small-scale producers – that participate in food production. Such changes in food production systems must also be re-embedded in consumption systems such that the environmental and social qualities inherent to the food consumed are valued – and not just financially.

The literature offering broad-brush solutions to the food crisis establishes that progress toward a more sustainable food system that meets the needs of the entirety of the global population requires more than change in individual behaviours and practices – there is also a need for change in governance structures. Suggestions for basic changes in food governance systems vary from those proposing more forward-looking policies for managing food stocks that avoid the sharp fluctuations in price, to calls for food policy that incorporates environmental and social well-being. These latter policies are necessary to promote consistent application and use of the practices required of individuals. There is, however, one drawback in all these solutions – which is also, perhaps, the most compelling explanation of the limited ability of alternatives to flourish against the totalizing claims of the business-as-usual model of neoliberalized global food governance. Individually, each of the solutions located at the level of food governance operates at specific scales and in extremely distinctive sites. In other words, they reside in a complex dilemma. They are all specifically scaled to solve the particular problems they are trying to address, yet collectively fail to cohere around a unified solution to global hunger with the power to motivate, enrol and discipline policy responses in the same manner achieved via the totalizing neoliberal solution.

We will go on to argue that mobilizing the notion of food utopias (or ideal food systems) has the potential to address aspects of this dilemma – unifying multiple, diverse and differentially-scaled governance solutions under a coherent set of wider values and ideas. Taking this pathway also aligns us with those who advocate that alternative governance systems and modes cannot be achieved through technocratic or policy processes and negotiations alone. Rather, we need a radical shift in our treatment of food, both in the values we attach to it as well as in our imaginings of more just and flexible systems. In other words, an alternative food utopia would serve to orient new governance systems by embracing flexibility and justice while recognizing the benefit of a unifying framework.

## Formulating responses 2: re-embedding food in sociocultural relations

In making the call to re-vision and recover food from its commoditized, abstracted forms we are aligning ourselves with a long tradition of social commentary and critique of market relations. Numerous social critics have already identified that the unintended consequences of commoditization contribute to, and maintain, the otherwise unjustifiable situations of excessive wealth and extreme poverty, and obesity and hunger that have become an accepted feature of global society. For example, Polanyi (1944) provided a scathing critique of the impact of market essentialism on social relations. He argued that much of the inequality experienced in 20th century society was the result of treating land and labour as commodities and, in the process, failing to fully value the environmental and social attributes they entailed. Further critique is offered by MacIntyre (1981) who argued that the transformation of goods into commodities could disassociate the product from the moral conditions of its production and consumption.

Several notable applications of critical perspectives on commoditization of food include those within the political economic tradition. Such analyses expose the impact of commoditization on consumers' awareness of the social and environmental impacts of the production of food. Butler and Dixon, in this book, refer to this critique in their discussion of the concept of 'metabolic rift'. The chapters by Campbell and McMichael use the framework of food regime theory to articulate how, at particular historical moments, food production and consumption patterns become aligned at a global level. The problem they identify lies in the current configuration of global food relations that is characterized by commoditization, cheapness and a relentless obscuring of the full character of food relations and the value created from food (and labour).

At a practical level, one response to these dynamics is apparent in the growing relevance and significance of alternative food systems – including fair trade, farmers' markets, organic and audited low-input production and designation of origin – for socially- and environmentally-conscious consumers. In academic analysis, these alternative food systems are identified as a potential means by which more equitable and sustainable pathways to global food security can be achieved (see Maye, *et al.*, 2007; Dowler, *et al.*, 2009). While these alternatives mobilize food cultures that contest the abstracted and commoditized norm, they can elicit ambiguous outcomes through their complex relationship with the ongoing operation of global market relations and consumer politics – as articulated by Smith and Lyons in this collection. Thus, alternative food systems can provide only partial solutions to the food crisis.

Another partial, and very important, element of change is presented in an edited book that documents the proceedings at a food security conference in Toronto

(Blay-Palmer 2010). The contributors to that book, including leading rural and food sociologists, emphasize the role of social movements at the consumption nexus of the food system in the reconfiguration of the global food system. Their emphasis is on responsible consumption and diets and involves the imagining of alternative food systems in which concerns for social justice and environmental sustainability become fundamental elements of consumption. Their approach resonates with the arguments of Lang *et al.*, (2009) for sustainable diets, albeit the latter authors envision a much more prominent and leading role for government policy. Lang and his co-authors promote the potential of 'choice editing' in the form of policies that preference sustainability relative to convenience and profit to reconfigure a pernicious food system. Their emphasis remains at a very rational and pragmatic level, however, focusing on the actions of consumers as individuals or as groups who express sustainable preferences in their consumption habits and demands for food qualities.

It is not only scholars in the social sciences who have arrived at this kind of understanding. Jules Pretty (2002), in his holistic analysis of the challenge of agricultural sustainability, provides a very cogent argument that a basic element of a change to more sustainable food production lies in reasserting the 'culture' in agriculture. He argues again in this volume that any kind of sustainable future 'means developing new and differing models, culturally-embedded and meaningful, that put food at the centre of economies and societies' (Pretty, this volume). Essential to these arguments is the recognition that society is intricately connected to ecological process.

The agreement that food needs to be re-embedded and revalued, as articulated by scholars from a range of perspectives, including its strong and repeated reinforcement in this collection, raises the question of how we can achieve this end. It is our contention that – in addition to the pragmatic changes to production and consumption of food – we, as a global society, need to establish a morally-informed perspective from which we can maintain a critical viewpoint on the ideology that perpetuates the existing food system. Put simply, we need a new utopian vision for food. Such a utopia, as noted by Stock and Carolan in this book, involves envisioning a society in which there is no hunger and which does not rely on reference to food as a quantity. Ultimately, it rests in the pursuit of a moral philosophy that recognizes the failure of market competition to ensure the equitable distribution of a product essential to human survival, such as food.

## Formulating responses 3: the potential power of Utopia

Up to this point we have offered that existing plans to 'feed the world' may be illusory sloganeering. Furthermore, we have categorized the recommendations that have been put forward in response to the food crisis into two broad categories (see Table 14.1): those that call for a tinkering of business-as-usual and those that re-imagine the global food system. Business-as-usual, while good business, does little to reduce real numbers of hungry people – as Claire Mahon emphasizes in this volume, 1.2 billion are hungry *now*. And the solution involves much more than redistribution, which merely reduces the economic realities to theoretical calculations. The question is not, 'How will (or even can) we feed the world?'; it is, more bluntly, 'Do we really want to?'

If the answer to this question is, 'Yes', then we must be able to envision an ideal food system that might accomplish that goal. We cannot confine our conception of a food-secure world to sufficient caloric intake. Such a goal represents a basement level of achievement that fuels population-related production arguments. The consistent message of this book is that such a narrow focus on production enables the persistence and normalization of inequality and abuse in the current food system, transforming food from the quintessential necessity for human life to the status of a tradable and quantifiable commodity. In the process we have lost our connection to food – to those who produce it, to those who bring it to market and prepare it, to those with whom we share it. Food too frequently becomes an excess, allowing us to select on the basis of appearance and popularity rather than embedded environmental and social qualities, and encouraging levels of waste beyond the capacity of the system. Business-as-usual – maintained by labour exploitation, environmental degradation, large numbers of starving and large amounts of waste – leads to polluted waterways, obesity epidemics, diabetes and heart disease outbreaks, topsoil erosion, lost biodiversity and on and on. As each of the contributors to this book has demonstrated, business-as-usual is irreparably broken. Yet, the idea that markets distribute goods in the most efficient and appropriate manner is elegantly simple and, therefore, hard to dislodge.

We must recognize that food is culturally, socially and politically instrumental to global societies. Invoking this recognition of a human right for sufficient and culturally appropriate calories, and of consumption as an element of human expression, can we imagine a global food system that balances issues of food as a human right, food sovereignty and global trade markets? The left-hand side of table 14.1 presents a catalogue of possible ways forward. A food utopia, as we envision it, can help us conceive of ways to get from where we are toward a more ideal food system in terms of justice and flexibility. We have to think the unthinkable, but in a way that provides a unified explanation with obtainable targets. As Ruth Levitas (2003) argues, by envisioning what the ideal might be – the utopian – we can better navigate away from what is not working. In proposing utopia as the foundation for achieving a more just and flexible food system, we utilize it in an atypical manner drawing inspiration from Max Weber, Ernst Bloch, Paul Ricoeur and Ruth Levitas. Our encouraging the creation of a more appropriate food utopia is not a futile exercise in seeking the impossible or proposing Pollyanna-ish solutions. Rather, we argue for utopia as a fundamental practice that enables us to imagine what is possible and to focus a critical perspective on the current food system. While the result is not an exact blueprint, it identifies a moral destination and provides a map *toward* a shared future in which food is redefined as an essential ingredient for healthy people and healthy communities at various scales.

By re-establishing food as a fundamental right and a building block to democracy, we can begin to do the same with our political institutions that have been hijacked in similar ways.

What, then, would an ideal global food system – a food utopia – look like? Revisiting the left-hand side of Table 14.1, we can propose a food utopia comprising a global food system that values communities, regions, cultural and ethnic diversity, the interdependence of social and environmental systems to produce and, more importantly, reproduce food and the biological conditions necessary for that production. It will provide a balance between local and regional flavours as well as the pathways to share those reciprocally. It will include plans for resilience such that local and international markets can draw on reserves when the inevitable drought or blight or agri-terrorist plot occurs. Floods will happen, temperature will fluctuate, farmer demography will change and corporations (at least for the immediate future) will play a huge role in how food is processed and distributed. Thus, an ideal food system would include more flexible, adaptive and just systems of production and governance, and more consumer options that respect cultural and social expression. Further, discussions about what an ideal food system might look like can provide important space to foster discussion between policymakers, agribusinesses, activists, academics, consumer groups and governments.

At the production level, the science of agroecology counters the hegemony of green revolution methods and the promotion of mono-cropping that erodes the inherent vibrancy and capacity for caloric output of farmland. Not only can agroecological methods contribute more than enough food, they also repair the rent social fabric by drawing on local knowledge systems and providing much needed employment opportunities (Altieri 2008; McIntyre, et al., 2009; de Schutter 2010). Research suggests we already produce enough food to feed the world and we can do more by re-imagining our assumptions about what is good agriculture (de Schutter 2010). Furthermore, the benefits of agroecology include more food, more biodiversity, more employment and more stability.

While calls for radical changes to the production methods and organizing principles of the global agriculture system have become more mainstream (as is evident in recent international and government sponsored research such as the IAASTD, the UN special rapporteur, the National Academies of Science from the US as well as the OFPS), opposition to business-as-usual – the quality side, as termed by Stock and Carolan – moves predominantly from the bottom up. Nor are the specific features of recommended changes for food governance new. The ideas of local self-provisioning, an emphasis on food sovereignty and livelihood preservation all speak to privileging the dignity of people as individuals and cultural groups. Decentralizing food systems speaks to issues about food miles, but also the cultural inheritance of food that underlies the growth of Slow Food and related movements. The 'food renaissance' also speaks to the importance of re-embedding food as a key glue for social relationships (Pretty 2002; Dowler, et al., 2009).

Our proposed food utopia can help bridge the two broad categories (incremental and transformational) of change identified in Table 14.1 and facilitate the engagement of their respective proponents in arriving at a solution to world hunger. Of course we can't discard the existing food system just because it is failing. There is much to be commended in terms of our ability to measure what is and is not working and these good things need to be retained. And there is no certainty that we will ever see the achievement of a perfectly responsible and fair global food system. But one of the great powers of imagination is envisioning where we might be headed – thinking the unthinkable helps move away from that which is broken. When you plan a holiday, typically you choose a place to go and then use a set of criteria (speed, comfort, aesthetic appeal) to decide on the best way to arrive at the travel destination. A utopia works in much the same way – we have to envision an ideal endpoint before we can even start on our way there. Further, a food utopia framework moves us away from the neoliberal assumptions that hide the inequality and failed promises behind the simplicity of its economic language. To properly feed the world, we must admit that too many suffer from hunger and that this situation is a problem worthy of our collective efforts. Hungry people are not a mere externality of a system that is yet to work perfectly.

Our contributors' chapters provide an incisive assessment of both the failings of business-as-usual and the remaining sources of hope in the existing food system. Our lack of appreciation of the socio-ecological nature of the food system (Pretty) encourages a focus on the economic and productive outcomes while ignoring the environmental (Lawrence et al.,) and social (Wald et al.,) costs of production. By valuing free trade despite its tremendous flaws (Pritchard, Lawrence et al.,), we ignore the problems it causes in local areas (Smith and Lyons). By conflating development (typically food aid) and economic growth (free trade) we create volatility in commodity prices as described in McMichael, and Neilson and Arifin. If we conceive of food as a human right (Mahon, Wald et al.,) and move away from food security as only a nutritional threshold (Butler and Dixon) we open up space for locally respondent creativity (Thornton). To reconceptualize an ideal food system is to re-envision aid and development, commodity and supply chains, alternative food systems and the relationships between them.

As a society do we believe that people are entitled to food, shelter, education and a right to livelihood; or do we value the freedom to accumulate wealth without guarantees against failure? (Is Monsanto too big to fail?) When the business of business became business and not livelihood (to paraphrase Calvin Coolidge) we assumed all would work out fine. But, as the IAASTD report stated: 'We cannot escape our predicament by simply continuing to rely on the aggregation of individual choices to achieve sustainable and equitable collective outcomes' (McIntyre, et al., 2009, p3).

Given that food is not a true commodity, there is a role for government regulation not only of price, but also of the movement and distribution of food. It is further apparent, however, that the complexity of the social and environmental relations of the global food system is beyond the capacity of governmental action.

The process of moving toward a more appropriate food utopia also requires an active citizenry as popularized by Tom Lyson's concept of civic agriculture (2004) and Melanie DuPuis and Sean Gillon's promotion of civic engagement as the basis for alternative agricultural systems (2009). Evidence of the potential of such action is found in the case of Toronto (Friedmann 2010). These examples represent new forms of food governance that, while more complicated and subjective than the idealized neoliberal market, we believe are a necessary element of a more just and flexible food system.

As Van der Ploeg (2009) points out, the EU is partially based upon the principle of subsidiarity, that is if something can be done lower down the scale, then it should be. This promotes a direct challenge to the neoliberal cul-de-sac. The Toronto example documented by Friedmann (2010) demonstrates a viable application of the principles of subsidiarity. Extended to the scale of a global food system, it is possible to envision an international federation of regional and local agreements, audit schemes and production goals that would be able to mediate the demands of stakeholders as diverse as Cargill, the raw milk producer in Madison, and the cassava grower in Thailand. At heart, the failure of the global food system is simply a symptom of a world trying to act as one and, only capable of governing as many. And it's the seductive potential of subsidiarity – to govern at appropriate scales – that fits so well with food.

On the whole, the framework of a food utopia provides us with the linguistic tools to incorporate a unifying theme, while offering pragmatic space for those who often conceive of one another as enemies, to engage in healthy debate. The neoliberal food agenda relies on the market to determine the nature and structure of the global food system. The common reference to market laws and forces implies, however, that a more just and flexible food system is too complicated and that it will sort itself out. The food crisis of 2006–2008 (which we are revisiting in 2011) is strong evidence that when things are left to sort themselves, a lot of people get rich and many more go hungry. The question is then, is this the kind of world we want to live in? The emerging consensus from experts and governments is, 'No, it's not; and here's what we can do about it'. Imagining an alternative food utopia is a first step. And it is an inclusive step. Instead of entrenching enemy positions, the collaborative imagining of an ideal global food system can foster much needed discussion. It is in such an imagining of a more just and flexible world of food – one that is appealing and acceptable to a broader segment of the global population – that we are most likely to get there.

## References

Altieri, M. A. (2008) *Small Farms as a Planetary Ecological Asset: Five Key Reasons Why We Should Support the Revitalization of Small Farms in the Global South*, Third World Network, Penang, Malaysia

Blay-Palmer, A., (ed) (2010) *Imagining Sustainable Food Systems: Theory and Practice*, Ashgate, Farnham, UK

Brown, L. (2011) *World on the Edge: How to Prevent Environmental and Economic Collapse*, W.W. Norton and Company, New York

Clapp, J. and Cohen, M. J. (eds) (2009) *The Global Food Crisis: Governance Challenges and Opportunities*, Wilfred Laurier University Press, Waterloo, Canada

Cribb, J. (2010) *The Coming Famine: The Global Food Crisis and What We Can Do to Avoid It*, University of California Press, Berkeley, CA

de Schutter, O. (2010) *Report Submitted by the Special Rapporteur on the Right to Food*, United Nations Human Rights Council, New York

Dowler, E., Kneafsey, M., Cox, R. and Holloway, L. (2009) '"Doing food differently": reconnecting biological and social relationships through care for food', *The Sociological Review*, vol 57, pp200–221

DuPuis, E. M. and Gillon, S. (2009) 'Alternative modes of governance: organic as civic engagement', *Agriculture and Human Values*, vol 26, pp43–56

Foresight (2011) *The Future of Food and Farming: Challenges and Choices for Global Sustainability*, Final project report, The Government Office for Science, London

Friedmann, H. 2010. 'Scaling Up: Bringing Public Institutions and Food Service Corporations into the Project for a Local Sustainable Food System in Ontario', pp157–172 in A. Blay-Palmer (ed) (2010) *Imagining Sustainable Food Systems: Theory and Practice*, Ashgate, Farnham, UK

Godfray, H. C. J., Crute, I. R., Haddad, L., Lawrence, D., Muir, J. F., Nisbett, N., Pretty, J., Robinson, S., Toulmin, C. and Whiteley, R. (2010) 'The future of the global food system', *Philosophical Transactions of the Royal Society B: Biological Sciences*, vol 365, pp2769–2777

Lang, T., Barling, D. and Caraher, M. (2009) Food Policy: *Integrating Health, Environment and Society*, Oxford University Press, Oxford

Lang,T. (2010) 'Crisis? What Crisis? The Normality of the Current Food Crisis', *Journal of Agrarian Change*, vol 10, pp87–97

Levitas, R. (2003) 'Introduction: The Elusive Idea of Utopia' *History of the Human Sciences*, vol 16, pp1–10

Lyson, T. A. (2004) *Civic Agriculture: Reconnecting Farm, Food, and Community*, Tufts University Press, Medford, MA

MacIntyre, A. (1981) *After Virtue*, Duckworth, London

McIntyre, B. D., Herren, H. R., Wakhungu, J. and Watson, R. T. (eds) (2009) *International assessment of agricultural knowledge, science and technology for development (IAASTD)*: synthesis report with executive summary: a synthesis of the global and sub-global IAASTD reports, Island Press, Washington, DC

Maye, D., Holloway, L. and Kneafsey, M. (eds) (2007) *Alternative Food Geographies: Representation and Practice*, Elsevier, Amsterdam

National Research Council (2010) *Toward Sustainable Agricultural Systems in the 21st Century*, The National Academies Press, Washington, DC

Polanyi, K. (1944) *The Great Transformation: The Political and Ecobnomic Origins of Our Time*, Beacon Press, Boston

Pretty, J. (2002) *Agri-Culture: Reconnecting People, Land and Nature*, Earthscan, London

Royal Society (2009) *Reaping the Benefits: Science and the Sustainable Intensification of Agriculture*, RS Policy Document 11/09, October 2009, The Royal Society, London

Sustainable Development Commission (2009) 'Setting the table: advice to government on priority elements of sustainable diets', Report to the Prime Minister, the first Ministers of Scotland and Wales and the First Minister and Deputy First Minister of Northern

Ireland, www.sd-commission.org.uk/publications/downloads/Setting_the_Table.pdf., accessed 4 April 2011, Sustainable Development Commission, London

UNEP (2011) 'Towards a green economy: pathways to sustainable development and poverty', www.unep.org/greeneconomy, accessed 4 April, 2011, United Nations Environment Programme, New York

Van der Ploeg, J. D. ( 2009) *The New Peasantries: Struggles for Autonomy and Sustainability in an Era of Empire and Globalization*, Earthscan, London

World Bank (2008) 'Rising food prices: policy options and World Bank response'. *Background note*, April 2008, World Bank, Washington, DC

# INDEX